Q
173
.W787
2000

KING   What

D0112701

SO MANY MYSTERIES, SO LITTLE TIME.
NOW YOU CAN UNRAVEL MOTHER NATURE'S
BEST-KEPT SECRETS WITH THE ASTOUNDING,
EYE-OPENING ANSWERS TO MODERN LIFE'S MOST
BAFFLING QUESTIONS.

- **How does a flame know which way is up?**

- **If the humidity gets to 100%, will I drown?**

- **Can a farmer really smell rain coming?**

- **Why do mirrors reverse left and right but not up and down?**

- **When my tires wear out, where has all the rubber gone?**

- **Why *isn't* it cold in space?**

Thanks to Robert L. Wolke, even the most complicated, unfathomable phenomena in our everyday world have clear, concise, well-I'll-be-darned explanations, from the weather to the food you eat to the reason your shower curtain seems to have a magnetic attraction to you. Discover the amazing unbreakable laws of science and Nature in the book that explains it all!

BOOKS BY ROBERT L. WOLKE

Impact: Science on Society

Chemistry Explained

What Einstein Didn't Know: Scientific Answers
    to Everyday Questions

What Einstein Told His Barber: More Scientific Answers
    to Everyday Questions

What Einstein Told His Barber

More Scientific Answers to Everyday Questions

ROBERT L. WOLKE

A Dell Trade Paperback

A DELL TRADE PAPERBACK

Published by
Dell Publishing
a division of
Random House, Inc.
1540 Broadway
New York, New York 10036

Cover design by Royce M. Becker

Copyright © 2000 by Robert L. Wolke

All rights reserved. No part of this book may be reproduced or transmitted in any form or by any means, electronic or mechanical, including photocopying, recording, or by any information storage and retrieval system, without the written permission of the Publisher, except where permitted by law.

Dell books may be purchased for business or promotional use or for special sales. For information please write to: Special Markets Department, Random House, Inc., 1540 Broadway, New York, N.Y. 10036.

DTP and the colophon are trademarks of Random House, Inc.

Library of Congress Cataloging in Publication Data

Wolke, Robert L.
    What Einstein told his barber: more scientific answers to everyday questions / Robert L. Wolke.
        p. cm.
    Includes index.
    ISBN 0-440-50879-7
    1. Science—Miscellanea. I. Title.
    Q173. W7875 2000
    500—dc21   99-045823

Printed in the United States of America

Published simultaneously in Canada

March 2000

10   9   8   7

BVG

I dedicate this book to my late father, Harry L. Wolke, to whom fate denied the opportunity of pursuing his own inclinations toward science and language, or even of seeing his son become a scientist and an author.

This one's for you, Pop.

JK. MARTIN LUTHER KING JK.
BRANCH LIBRARY
3436-42 SOUTH KING DRIVE
CHICAGO. ILLINOIS 60616

Acknowledgments

I want to express my dying gratitude to all of my friends who said, "Hey, Bob, I thought of a great question for your book the other day, but I forgot it."

I earnestly thank two nice guys with whom it has been a pleasure to work: my agent, Ethan Ellenberg, and my editor, Mike Shohl. Ethan skillfully navigated my proposal through the "shohls" of contract negotiation, while Mike wielded his blue pencil with understanding and restraint, getting all my jokes and allowing them to survive.

Diana Zourelias's delightful drawings add an even lighter touch to what would otherwise have been a relentlessly gray-paged volume. All I did was give her a paragraph describing each situation to be illustrated, and her whimsical creativity took it from there.

I am indebted to Richard E. Eckels for guiding me to a true explanation of how airplanes fly.

The two women in my life, my daughter, Leslie, and my wife, Marlene, never flagged in their encouragement or in their regard for my work despite my computer's often competing with them for my time. For that and for their love and admiration I am grateful every day of my life.

Contents

Hot Stuff   **77**

If you can't stand the heat, get out of the universe
. . . because heat is the ultimate form of energy.

Is 100 degrees twice as hot as 50 degrees? What is
temperature, anyway? How cold can it get? Why is the bathroom floor so cold on your bare feet? How hot can it get?
How does a flame know which way is up? Why is a candle
flame tapered at the top? Could we counteract global warming by turning on all our air conditioners? What's so dangerous about high voltage? Why doesn't it rain dead sparrows?
. . . And more.

The Earth Beneath Our Feet   **103**

**O happy Earth, whereon thy innocent feet do
ever tread!—SPENSER**
. . . And thy innocent mind doth ever strive to understand.

Why does Earth pull everything toward its exact center? How
does hot air defy gravity by rising? If hot air rises, why is it
colder in the mountains? Does it ever get too cold to snow?
If Earth is spinning so fast, why don't we fly off? Can the
astronauts see Earth turning beneath them? Would a polar
bear weigh less at the equator? Do toilets flush counterclockwise in the northern hemisphere and clockwise in the southern hemisphere? Can you stand an egg on end during the
vernal equinox? Why is nuclear energy unique on Earth?
How does radiocarbon dating work? . . . And more.

5 Heavens Above!   142

These earthly godfathers of heaven's lights
    That give a name to every fixed star
Have no more profit of their shining nights
    Than those who walk and know not what they are.
—SHAKESPEARE

Sorry to disagree with you, Will, but it's much more fun if you know what they are. The air, the sky, the moon and the stars are all up there for us to comprehend.

How do odors find your nose? Can you operate a vacuum cleaner in a vacuum? Why does a lion tamer's whip make such a loud "crack"? What is the sound barrier made of? Why does thunder sound the way it does? Why is the moon so much bigger when it's near the horizon? Why do the stars twinkle? How does the moon keep one side always facing Earth? How do the oceans' tides work? Does the moon ever turn blue? Why is it cold in space—or is it? . . . And more.

6 All Wet   178

You can lead a horse to water, but you can't make him think.—WOLKE

We humans, however, can ponder the remarkable properties of the most abundant chemical on Earth.

What color is water? Why are the oceans blue? And salty? Precisely where is sea level? Why does spilled coffee dry to a ring? Why does your shower curtain cling to you? Where do your socks go when they disappear in the laundry? What's the most expensive ingredient in laundry detergents? (Advertising.) Is glass a liquid? What makes ice cubes cloudy? If the humidity got to be 100 percent, would we drown? How can you clear your fogged-up windshield? Can a farmer smell rain? . . . And more.

7 **Stuff and Things**   **215**

He who dies with the most stuff wins.
—YUPPIE PHILOSOPHY
We have surrounded ourselves with hundreds of material things that we use but may not really understand.

Are airplanes safe? How does an eraser erase? Why does rubber stretch? Why are cars noisy? Why do clothes wrinkle? How does a skateboard work? What happens when you shake a bottle of soda? Can you get electricity out of a lemon? Are smoke alarms radioactive? Can fertilizer explode? How do sailors keep clean? If you stopped dusting, how long would it take to be buried in dust? Can you unburn a match? . . . And more.

Introduction

I know what you're thinking. You're thinking, "Did Einstein even *have* a barber?"

You've seen his pictures, right? And it's perfectly clear that the great man devoted a lot more time to cultivating the inside of his head than the outside.

But this book isn't about barbers, and it isn't even much about Einstein. (His name comes up only four times.) It is a book of scientific small talk, the kinds of things that Einstein *might* have talked about with his barber—simple things that may have been trivial to the great scientist, but that the rest of us may wonder about.

There are many science-is-fun books for young readers. But it isn't only children who wonder "Why?" or "How?" Curiosity doesn't end at puberty, nor does the genuine fun of understanding why things happen. And yet, once we are "done with science" in school we encounter few books for people of any age who are simply curious about their everyday surroundings and derive pleasure from knowing what makes them tick. This is that kind of book.

Maybe you are convinced that science is "not for you," that it is inherently difficult stuff, and that if you were to ask a question the answer would be too technical and complicated for you to understand. So you just don't ask. You may have come to these conclusions because of unfortunate experiences with school science classes or simply from the science stories in newspapers and magazines and on television. These stories are by their very nature guaranteed to be technical and complicated, because they are about the latest discoveries of leading scientists. If they weren't, they wouldn't be news. You won't see a TV special on why the bathroom floor feels so cold on your bare feet. But the explanation of that phenomenon (see p. 83) is science, every bit as much as a discussion of quarks or neutron stars.

Science is everything you see, hear and feel, and you don't have to be an Einstein or even a scientist to wonder *why* you are seeing, hearing and feeling those things, because in most cases the explanations are surprisingly simple and even fun.

This is not a book of facts. You will not find answers here to questions such as "Who discovered . . . ?" "What is the biggest . . . ?" "How many . . . are there?" or "What is a . . . ?" Those aren't the kinds of things that real people wonder about. Collections of answers to such contrived questions may help you win a trivia contest, but they are not satisfying; they don't contribute to the joy of understanding. The joy and the fun come not from mere statements of fact but from *explanations*—explanations in plain, everyday language that make you say, "Wow! Is *that* all there is to it?"

There are well over a hundred explicit questions addressed in this book, but that by no means limits the number of things that are actually explained. The physical world is a complex web of goings-on, and nothing happens for a single, facile reason. In science, every answer uncovers new questions, and no explanation can ever be complete.

Nevertheless, I have written each question-answer unit to be self-contained, to be read and understood independently of all the others. This must inevitably lead to some overlap— an essential link in logic cannot be omitted simply because it is treated in greater detail elsewhere. But as every teacher knows, a bit of repetition never hurt the learning process.

Whenever another Q&A unit contains closely related information, you will be referred to the page number on which that unit *appears*. Thus, there is no need to read the book sequentially. Read any unit that catches your eye at any time. But don't be surprised if you are lured into a web of related units by the page references. Follow the lark. That way, you'll be following trains of thought sequentially, as if they had been laid out in a (heaven forbid) textbook, which neither of us wants. You've been there, and I've done that. And whenever a complete explanation requires a little more detail than you may be in the mood for, that detail is banished to a Nitpicker's Corner. There, you may either continue reading or just skip it and move on to another question. Your call.

I have studiously avoided using scientific terms. I believe that any concept that is capable of being understood should be explainable in ordinary language; that's what language was invented for. But for their own convenience, scientists use linguistic shortcuts that I call "Techspeak." When a Techspeak word is inescapable, or when it is a word that you may have heard and avoiding it might seem contrived, I define it in plain language on the spot. You will find the definitions of some useful Techspeak words in the back of the book.

I assume no previous scientific knowledge on your part. There are three ubiquitous Techspeak words, however, that I use without taking the trouble to define each time: atom, molecule and electron. If you're a bit skittish about your familiarity with them, check them out in the Techspeak list before you begin.

Scattered throughout the book you will find a number of Try Its—fun things that you can do in your own home to illustrate the principles being explained. You will also find a number of Bar Bets that may or may not win you a round of drinks, but that will certainly get a spirited discussion going.

When Albert Einstein was in residence at the Institute for Advanced Study at Princeton University, an eager young newspaper reporter approached him one day, notebook in hand. "Well, Professor Einstein," he asked, "what's new in science?"

Einstein looked at him with his deep, soft eyes and replied, "Oh? Have you already written about all the *old* science?"

What he meant was that science isn't to be characterized only by the latest headline-making discovery. Scientific observation has been going on for centuries, and in that time we have learned a tremendous amount about the world around us. There is a vast heritage of knowledge that explains ordinary, familiar happenings.

That's the "old science." Everyday science. That's what this book is about.

Movin' and Shakin'

Everything is moving.

You may be sitting quietly in your armchair, but you are far from motionless. I don't mean merely that your heart is beating, your blood is coursing through your veins and you are panting at the prospect of learning so many fascinating things from this book. In short, I don't mean simply that you are physically and mentally alive.

I mean that while you are sitting there so peacefully, Earth beneath your feet is spinning you around at about 1,000 miles per hour (1,600 kilometers per hour). (The exact speed depends on where you live; see p. 119). Mother Earth is simultaneously hauling you around the sun at 66,600 miles per hour (107,000 kilometers per hour). Not to mention the fact that the solar system and all the stars and galaxies in the universe are racing madly away from one another in all directions at incredible speeds.

Okay, you knew all that. Except maybe for the exact speeds. But we're still not done.

You are made of molecules. (Yes, even you.) And all your molecules are vibrating and jiggling around to beat the band, assuming that your body temperature is somewhere above absolute zero (see p. 82). In motion also are many of the atoms of which your molecules are made, and the electrons of which the atoms are made, and the electrons, atoms and molecules of everything else in the universe. They were all set into motion about 12 billion years ago (see p. 175) and have been quivering ever since.

So what *is* motion? In this chapter we'll see how everything from horses to speeding automobiles, sound waves, bullets, airplanes and orbiting satellites move from one place to another.

Horsing Around on the Highway

Why do they drive on the left in some countries and on the right in others?

It goes back to the fact that most humans are right-handed.

Long before we had modern weapons such as guns and automobiles, people had to do battle using swords and horses. Now if you are right-handed, you wear your sword on the left, so that you can draw it out rapidly with your right hand. But with that long, dangling scabbard encumbering your left side, the only way you can mount a horse is by throwing your free right leg over him. And unless you are in a Mel Brooks movie and want to wind up sitting backward on your steed, that means that the horse's head has to be pointing to your left. To this day we still train horses to be saddled and mounted from their left sides.

Now that you are mounted, you will want to stay on the left side as you start down the road, because anyone coming toward you will be on your right, and if that someone turns out to be an enemy, you can whip out your sword with your right hand and be in position to run the scoundrel through. Thus, prudent horsemen have always ridden on the left side of the road.

This left-side convention was also honored by horse-drawn carriages in order to avoid annoying collisions with horsemen. When horseless carriages made their appearance, some countries continued the habit, especially during the overlap period when both kinds of carriages were competing for road space.

So why do people drive on the right in the U.S. and many other countries?

When swords went the way of bows and arrows, the need for defending one's right flank disappeared and traffic rules were suddenly up for grabs. Younger or less tradition-bound countries migrated to the right, apparently because the right-handed majority feels more comfortable hugging the right side of the road. It quickly occurred to left-handed people that it was unhealthy to argue with them.

Some countries that I've been in must have large populations of ambidextrous people, because they seem to prefer the middle of the road.

Four-Grief Clovers

Why do highway and freeway intersections have to be so complicated, with all those loops and ramps?

They enhance the traffic flow—from construction companies to politicians' campaign chests.

Sorry.

They allow us to make left turns without getting killed by oncoming traffic. It's a matter of simple geometry.

When freeways and superhighways began to be built, engineers had to figure out how to allow traffic to make turns from one highway to an intersecting one without stopping for red lights. Because we drive on the right-hand side of the road in the U.S., right turns are no problem; you just veer off onto an exit ramp. But a left turn involves crossing over the lanes of opposing traffic, and that can cause conflicts that are better imagined than expressed.

Enter the cloverleaf. It allows you to turn 90 degrees to the left by turning 270 degrees to the right.

Think about it. A full circle is 360 degrees; a 360-degree turn would take you right back to your original direction. If two highways intersect at right angles, a left turn means turning 90 degrees to the left. But you'd get the same result by making three right turns of 90 degrees each. It's the same as when you want to turn left in the city and encounter a "No Left Turn" sign. What do you do? You make three right turns around the next block. That's what the loop of a cloverleaf does; it takes you 270 degrees around three-quarters of a circle, guiding you either over or under the opposing lanes of traffic as necessary.

The highway interchange is a *four*-leaf clover, rather than a two- or three-, because there are four different directions of traffic—going, for example, north, east, south and west—and each of them needs a way to make a left turn.

For readers in Britain, Japan and other countries where they drive on the left, just interchange the words "left" and "right" in the preceding paragraphs, and everything will come out all right. That is, all left. You know what I mean.

Ready, Set . . . Jump!

If every person in China climbed to the top of a six-foot (two-meter) ladder and then all jumped off at the same time, could it nudge Earth into a different orbit?

No, but it sure would create a windfall for Chinese podiatrists.

I suppose that everybody picks on China when they ask this question because China is the most populous country on Earth, containing 2.5 billion potentially sore feet.

There are really two questions here, aside from the question of why people who ask this question don't have anything better to do. (Just kidding; it's fun to wonder about such things.) The first question is how strong the jump-thump would be, and the second question is whether any size thump at all could change Earth's orbit.

It's easy to calculate the amount of energy from a gravitational fall. (And don't tell me they're not falling because China is upside down.) Assuming a population of 1.2 billion Chinese weighing an average of 150 pounds (68 kilograms) each, their collective pounce would hit the ground with an energy of 1.6 trillion joules. (A joule is just a unit of energy; don't sweat it.) That's just about the amount of energy released in a medium-sized earthquake measuring 5.0 on the Richter scale. Such earthquakes have been occurring for millions of years, and there is no evidence that they have nudged Earth into different orbits.

But no amount of earthquake or footquake energy could change the orbit anyway, so both earthquakes and Chinese ladders are irrelevant. Planet Earth continues circling the sun because it has a certain amount of momentum, which means that it has a certain amount of mass and a certain velocity, because momentum is a combination of mass and velocity. Our planet carries along with it everything that is attached to it by gravity, including jumping Chinese and acrobats on trampolines. We're all one big package of mass, and no amount of jumping up and down can change Earth's total amount of mass. Nor can it change the planet's velocity, because all the Chinese are being carried along through space at the same speed as the rest of the planet; we're all in one big, interconnected spaceship. You can't change the speed of your car by pushing on the windshield, can you? Nor can you lift it by pushing on the inside of the roof.

We might put it in terms of Newton's Third Law of Motion, which you must have heard a million times (and will again, if I have anything to do with it): "For every action there is an equal and opposite reaction." Push on a brick wall and the wall pushes back. If it didn't, your hand would go straight through. When the Chinese land, their feet hit the ground with a certain amount of force, but at the same time the ground hits their feet with an equal amount of force in the opposite direction. Thus, (a) there is no net (unbalanced) force that could affect our planet's motion and (b) their feet hurt.

Jump . . . Now!

If I'm in an elevator and it starts to fall to the bottom of the shaft, can I jump up at the last instant and cancel the impact?

Ho hum. I don't know how many times this question has flashed into the minds of worrywarts in elevators, or how many times it has been asked of every friendly neighborhood physicist. It is easy to answer in one word (No), but thinking about it does raise a whole bunch of fun questions.

First, here's the quick answer: Your objective is to arrive at the bottom of the shaft like a feather, without any appreciable downward speed, right? That means that you have to counteract the elevator's downward speed by jumping upward with an equal amount of speed. The elevator (and you) might be falling at, say, 50 miles per hour (80 kilometers per hour). Can you jump upward with anywhere near that speed? The best basketball players can jump at maybe 5 miles per hour (8 kilometers per hour). End of quick answer.

Let's consider the instant before your elevator's cable snaps.

In the seventeenth century, long before elevators, Sir Isaac Newton (1642–1727) realized that when a body exerts a force on another body, the second body exerts an equal and opposite force on the first body. Today, that's known as Newton's Third Law of Motion. When you're standing on the elevator floor and gravity (force number one) is pulling you down against the floor, the floor is pushing you back up with an equal force (force number two). That's why gravity doesn't win out and make you fall down the shaft. It's the same with the elevator car itself; in this case it's the cable's upward pull that counteracts gravity's downward pull on the car. So neither you nor the elevator falls down the shaft. You both move upward or downward at a speed that is controlled by a motor's slow winding and unwinding of the cable from a big drum at the top of the shaft.

When the cable snaps, both the upward pull of the cable and the upward push of the floor are suddenly gone, so both you and the elevator are free to succumb to gravity's will and you both begin to fall. For an instant you are left floating—feeling "weightless" because the customary push of the floor on your feet is gone. But following that instant of blissful suspension, gravity has its way with you and you fall, along with the elevator.

NITPICKER'S CORNER

About that moment of "weightlessness" when the elevator begins to fall: Obviously, you haven't really lost weight. Earth's gravity is still pulling on you as it always has, and the strength of that pull is what we call weight (see p. 23). What you've lost is *apparent weight*. Your weight just isn't apparent because you're not standing on a scale or a floor that feels your pressure and presses back upon your feet.

Of course, this whole question of falling elevators is hypothetical because elevator cables just don't snap. And even if they did, there are spring-loaded safety devices that would keep the car from falling more than a couple of feet. But, as roller coasters prove, some people seem to enjoy the contemplation of imminent disaster.

If you happen to be one of those roller coaster fans, that "floating" feeling you get as the car falls from one of its high spots is exactly the same thing you'd feel in a falling elevator. It's called *free fall* (see p. 26). Astronauts in orbiting spacecraft also feel it (see p. 23).

Dead Tread

When my car's tire treads wear out, where has all the rubber gone?

It has been rubbed off—and no, that's not why they call it rubber (see p. 217)—onto the road, whence it was scattered in the form of fine dust into that vast, complex everywhere that we call the environment. Some of it was then washed off the road and into sewers by rain, or else it was blown around by the wind and eventually fell or was rained out of the air onto any and all surfaces. Eventually, all the rubber joined the soil and the seas as part of the Earth from which it was born. Like everything else, a dead tread returns unto dust.

We tend to think of automobile tires as rolling smoothly along, without any scuffing against the road that might scrape away rubber. That could be true only if there were no resistance whatsoever between the tire's surfaces and the road's surface. And if there were no resistance, your tires couldn't get a grip and you'd go nowhere. You'd get a spectacular warranty on a set of tires like that, because they'd never wear out.

Between any two surfaces that are attempting to move past each other—even a tire and a road—there is always some resistance; it's called friction. Even rolling wheels experience friction against the road, although rolling friction is a lot less than sliding friction. When necessary, you can roll your car straight ahead by pushing, but just try to slide it sideways.

Friction gobbles up some of the energy of motion and spits it out as heat. If there were no diminishment of motion by the conversion of some of it to frictional heat, a machine could go on forever without slowing down: perpetual motion. Because there always must be some frictional heat loss, however small, every device that has ever been touted as a perpetual motion machine has to be a fake, however well-intentioned its inventor.

TRY IT   If you don't think that tire-against-road friction makes heat, just feel your tires before and after driving for an hour or so on the freeway. Much of the heat you'll feel comes from friction against the road, but some comes also from the continual flexing and unflexing of the rubber (see p. 219).

Regarding the disappearing tread on your tires: Wherever there is frictional resistance between two materials, one of them has to "give"—that is, have some of its molecules scraped off by the other. Between your soft tire and the hard highway, it's no contest; it's the rubber that gives and gets rubbed off gradually in tiny particles.

If all of our roads were made of a substance that is softer than rubber, the roads would wear out instead of the tires. Instead, our society has decided that it's less trouble for car owners to replace their tires than for governments to continually replace road surfaces. Then why, you may ask, do we continually have to dance the orange barrel polka to get through interminable road reconstruction zones? Unfortunately, I can answer only scientific questions, not political ones.

The squealing tires in movie car chases are the result of sliding friction: rubber scraping, rather than rolling, on the pavement. On a microscopic scale, we would see the tire alternately grabbing and slipping thousands of times per second, producing a series of chattering vibrations that fall in the frequency range of a screech. It's easy to see that with all of this frictional dragging of rubber against the road, a lot of

rubber will be rubbed off. In fact, the friction makes enough heat to melt some of the rubber, which paints itself onto the road as a black skid mark.

You Didn't Ask, but . . .

Why are the tires on racing cars so smooth? You'd think they'd need all the traction they could get.

That's precisely why they're smooth. Regular tires waste a lot of their potential road-grabbing surface by having grooves, which act like gullies to channel out rain and mud. But racing cars usually compete in good weather, so the rain-and-mud grooves aren't necessary. They're just wasted space that can better be used to add more road-grabbing rubber for better handling in turns and better braking response. To get even more road-grabbing surface, the tires are made much wider than those on your family chariot. And they're made of a softer rubber that wears off like crazy onto the track. You think *you* don't get good tire mileage? Why do you think they're always stopping to change tires?

Ready, Aim, Scram!

In movie westerns, and even in many parts of the world today, people fire guns straight up into the air as warning shots or just to make noise during a celebration. But those bullets have to come down somewhere. How dangerous will they be if they hit somebody?

Quite dangerous. As we'll see, physics tells us that when it hits the ground the bullet will have the same velocity it had when it left the muzzle of the pistol, which can be 700 to 800 miles per hour (1,100 to 1,300 kilometers per hour). But that ignores air resistance. More realistically, the bullet's landing speed can be around 100 to 150 miles per hour (160 to 240 kilometers per hour). That's fast enough to penetrate human skin, and even if it doesn't penetrate it can still do a lot of damage. But just try to tell that to the idiots who like to shoot their guns "harmlessly" into the air.

There are two kinds of forces that affect the bullet's speed on the way up and on the way down: gravity and air resistance. Let's look first at the effects of gravity, neglecting air resistance entirely.

It will be easier to understand the bullet's flight if we consider it in reverse. That is, we'll start at the instant at which the bullet has reached the top of its flight and is just starting to fall downward. Then we'll consider its upward journey and compare the two.

Gravity is a force that operates on a falling object—and is indeed what makes it fall—by pulling on it, attracting it toward the center of the Earth (see p. 104), a direction that we call "down." As long as the object is in the air, gravity keeps on tugging on it, urging it to fall faster and faster. The longer it falls, the more time gravity has to work on it, so the faster it falls. (Techspeak: It *accelerates*.)

The strength of Earth's gravitational field is such that for every second of pull—that is, for every second that an object is falling—the object speeds up by an additional 32 feet per second (9.8 meters per second) or about 22 miles per hour (35 kilometers per hour). It doesn't matter what the object is or how heavy it is, because the strength of the gravitational field is purely a characteristic of Earth itself. So for every second of downward fall, the bullet gains 22 miles per hour (35 kilometers per hour) of speed. If it falls for ten seconds, its speed will be 220 miles per hour (350 kilometers per hour), and so on.

But gravity was pulling on the bullet with the same force when it was on its way up. That's what slowed it down so much that it eventually reached zero speed at the top of its flight before starting to fall. For every second that it was on its way up, gravity's pull *removed* 22 miles per hour (35 kilometers per hour) of speed. The total amount of speed removed on the way up must be the same as the total amount of speed regained on the way down, because the gravitational effect was the same all the time. If that weren't true, the bullet would have to have acquired some speed or lost some speed because of some other outside force. And there was no other outside force (except air resistance, and we'll get to that).

So we see that what gravity taketh away on the way up, gravity giveth back on the way down. On the basis of gravitational effects alone, then, the bullet would have no more or less speed when it hits the ground than it had when it left the gun: its muzzle velocity, and that's how fast it will be going when it hits the ground.

. . . Or an innocent bystander.

Up to now, we've ignored the slowing-down effect of the air. As you can tell by sticking your hand out the window of a moving car, the faster you go the more the air tries to hold you back. So as our bullet falls faster and faster under the influence of gravity, air resistance tries to make it go slower and slower. Pretty soon, the two conflicting forces become equal and cancel each other out. After that, no matter how much farther the object falls it won't go any faster. It has reached what physicists like to call its *terminal velocity,* which is Techspeak for final speed.

(Because "terminal velocity" is such an impressive-sounding term, many an innocent physics student—I was one—gets the impression that it's some kind of fundamental limitation of Nature, like the speed of light. But there's absolutely nothing sacred or fixed about it. The final speed of a falling object simply depends on its size and shape, and on how it catches the air. If you fall out of an airplane, your terminal velocity will certainly be a lot less if you're wearing a parachute. Teams of sky divers adjust their air resistance by making their bodies more compact or more extended, so they can rendezvous at the same terminal velocity and frolic around together before pulling their rip cords.)

If a shooter is fairly close to a target, there isn't much opportunity for air resistance to slow the bullet down during its short flight. Even when fired into the air, a streamlined object like a bullet doesn't suffer much air resistance on the way up, because it is pointing straight ahead along its path. But during its fall it is probably tumbling, or even more likely falling base-first, because that's the most stable orientation for a bullet-shaped object. The air resistance on a tumbling or base-first bullet is quite a bit greater than on a straight flyer, so it may be slowed down substantially on the way down and end up quite a bit slower than its muzzle velocity. One expert estimates that a .22LR bullet with a muzzle velocity of 857 miles per hour (1,380 kilometers per hour) might fall to the ground with a velocity of 96 to 134 miles per hour (154 to 216 kilometers per hour), depending on how it tumbles. That's more than enough speed to do serious or lethal damage to a cranial landing site.

And by the way, the jerk who fires the bullet isn't very likely to be hit by it, no matter how carefully he aims straight up. In one experiment, out of five hundred .30-caliber machine-gun bullets fired straight upward, only four landed within 10 square feet (3 square meters) of the gun. Wind has a great effect, especially since .22- to .30-caliber bullets can reach altitudes of 4,000 to 8,000 feet (1,200 to 2,400 meters) before falling back down.

War Is . . . Swell

Why do guns put spin on their bullets?

A spinning bullet flies farther and truer than it would without the spin. And if your favorite sport is football rather than shooting, just about everything I'm going to say about spinning bullets also goes for spiraling passes.

The fact that a spinning bullet or football goes farther may sound strange, because you'd think that the range would depend only on the amount of energy it gets from the gunpowder charge or the quarterback's arm. But bullets and footballs have to fly through the air, and air drag plays an important part in any projectile's trajectory, whether it is fired from a handgun, rifle, machine gun, howitzer or arm.

First, let's see how a gun makes the bullet spin.

Running the length of the inside of the barrel are spiraling grooves, called rifling. As the bullet passes through the barrel, these grooves cut into it, making it rotate to conform to the spiral. Some guns have grooves that twist to the right and some have grooves that twist to the left; it doesn't matter. (And no, they don't twist one way in the northern hemisphere and the other way in the southern hemisphere. But see p. 125.)

Early bullets were round balls of lead, like miniature cannonballs. Bullet-shaped (Techspeak: *cylindroconoidal*) bullets were developed around 1825, when it was found that they maintained their speed better in flight. That's because for a given weight of lead an elongated, tapered-nose shape meets with less air resistance than a round ball; it's streamlined.

But there's a problem with elongated bullets that spherical bullets don't have. When an elongated bullet is fired, any tiny irregularities on its surface can catch the air and push it slightly sideways, so that its nose is no longer pointing straight ahead. This slight misalignment increases the air resistance on the forward side, which turns the bullet even more. Pretty soon it is tumbling end-over-end, which causes even more air drag, seriously shortening its range and pushing it off-course. Thus, both distance and accuracy suffer.

That's where the rifling comes in. If the bullet is spinning properly around its long axis as it flies, it resists any change in its orientation or direction of flight. The reason for that is that a heavy, spinning object has a lot of momentum. Not only does it have momentum along its direction of travel (linear momentum), but because of its spin it also has rotational momentum, or what physicists call angular momentum. And momentum, whether linear or angular, is hard to upset. In fact, the momentum of an object will remain unchanged unless and until it is disturbed by some outside force. (Techspeak: *Momentum is conserved.*) The spinning bullet, therefore, will maintain its angular momentum by spinning with its axis in the same direction for as long as it is in the air, because there is no outside force to disturb it. Those tiny surface irregularities are now peanuts compared with the bullet's substantial amount of angular momentum.

With its nose pointed straight ahead, the projectile encounters less air resistance and thus flies farther and truer. When it ultimately hits an object, its momentum—both linear and angular—still won't disappear, but will be transferred to the unfortunate target—or in the case of a football, the fortunate receiver.

International law actually requires that bullets spin. Otherwise, a tumbling bullet might hit its victim sideways, doing more damage than if it had made a nice, clean, round hole. It's just one of those niceties of war: If you're going to kill somebody, please do it neatly.

The Geneva Convention spells out certain other niceties about how to kill people. For example, because lead is soft and deformable, it can go splat when it hits its target, again producing a very unsightly hole. So bullets have to be jacketed with a harder metal, such as copper. The world's military establishments gladly comply with that requirement, but it's not because of any humanitarian motives. It's because modern military assault weapons fire their bullets at such high speeds that if they weren't jacketed with high-melting copper the lead would melt from friction with the air, making them fly erratically and miss their targets. After all, a clean, round hole in an enemy is so much preferable to no hole at all.

You Didn't Ask, but . . .

Why does the Lone Ranger use silver bullets?

They serve mostly as a calling card, but they do have a very slight advantage over lead.

Ordinary bullets are made of lead because lead is so heavy, or dense. And it's cheap. We want a bullet to be as heavy as possible because we want it to have as much damage-causing energy as possible when it hits its target, and energy is a combination of mass and speed. (Techspeak: *Kinetic energy is directly proportional to the mass and to the square of the velocity.*) It's easier to gain energy by increasing the bullet's mass than by increasing its velocity, because increasing the velocity would require a longer barrel in order to give the explosion's gases more time to accelerate the bullet.

A silver bullet is about 7.5 percent lighter than a lead bullet of the same length and caliber. Since a given powder charge imparts the same amount of energy to any bullet, the lighter silver bullet must travel faster. It works out to be 4 percent faster than a lead bullet.

So the Lone Ranger's silver bullets get to their targets very slightly sooner than a lead bullet would. If the bullet's velocity is 1,000 feet per second (300 meters per second) and an outlaw fifty feet (fifteen meters) away is drawing his gun, the silver bullet gives our hero a two-millisecond advantage—not even long enough for Tonto to say, "Ugh!"

Also, because silver is a lot harder than lead, when the Lone Ranger shoots the gun out of a bad guy's hand—he never shoots the guy himself—it must really sting. And when it strikes, instead of the dull thud of lead, a silver bullet makes a great "ping" sound for the microphones that seem always to be nearby.

BAR BET The Lone Ranger's silver bullets fly faster than lead bullets.

How to Stop an Airplane

When there's an airplane flying overhead, why is it that when I walk in the opposite direction it looks as if it's almost stationary? Certainly my walking speed is peanuts compared with the plane's speed, so how can it be having any effect?

Whether we realize it or not, we judge the motion of an airplane in the sky by its relation to common things on the ground, such as trees, telephone poles and houses. That's the only way motion can be detected: in relation to something else. There's no such thing as absolute motion; it's all relative to something else (see p. 119). So the faster the plane appears to be passing the trees and houses, the faster we judge the plane to be moving.

But when you yourself are moving in relation to the trees and houses, you upset this simple association because the trees and houses appear to be moving also. As you walk forward, they appear to be moving backward, don't they? Of course, you know that they're not *really* moving backward because your daddy told you so when you were two years old.

So as you walk forward (which, I trust, is your customary direction of locomotion), but in the *opposite* direction from the airplane's, the trees and houses also appear to be moving backward with respect to your direction; that is, they appear to move *in the same direction as the plane.* It appears, then, that the airplane and the houses are moving together; the plane doesn't seem to be overtaking them. And any airplane that can't even pass a house would seem to be one very slow airplane.

Want to do the passengers a favor and get them to their destination sooner? Just walk in the same direction as the plane. As the trees and houses "move backward" it'll look as if the plane is passing them even faster.

It's Truly Not Bernoulli

I just can't bring myself to believe that huge airplanes can fly, supported as they are on thin air. How do they do it?

Join the club. Even though I know something about how airplane flight works (and you will too, soon), it never ceases to amaze me. I remember landing after a transatlantic flight in a Boeing 747 and being directed by the crew to deplane directly onto the ground and into a waiting bus, instead of through one of those people tubes. I looked up in utter disbelief at the four-hundred-ton monster that had just wafted me across the Atlantic Ocean at an altitude of more than five miles above Earth's surface.

My awe was magnified by the fact that back when I was "taught" what makes airplanes fly, I was misled. In spite of the fact that most flight training manuals attribute an airplane's lift to something called Bernoulli's Principle, that is *not* the main reason airplanes stay up. It just happens to be a quick, easy explanation, but like all simple answers it is misleading, bordering on downright wrong.[1]

First, let's put the Swiss mathematician Daniel Bernoulli (1700–1782) on the witness stand and see what he has to say for himself.

In 1738 Bernoulli discovered that as the speed of a moving fluid (gas or liquid) increases, its pressure on adjacent surfaces decreases. For example, air that is blowing by as a horizontal wind doesn't have the time or energy, so to speak, to press very hard upon the ground.

How does this affect airplanes?

The top surface of a conventional airplane wing is humped upward, while the bottom surface is relatively flat.

---

[1]The following treatment of airplane flight is based upon David Anderson and Scott Eberhardt's article "How Airplanes Fly: A Physical Description of Lift" (*Sport Aviation*, February 1999), which was pointed out to me by Richard E. Eckels.

As the plane flies, air sweeps over these two surfaces. On its way to the back (trailing) edge of the wing, the air on the top surface has farther to go because of its curved path. The Bernoulli-Makes-Planes-Fly advocates claim that the top and bottom air must reach the wing's back edge at the same time—that's called the *equal transit time* assumption—and that inasmuch as the top air has farther to travel it must move faster. According to Mr. Bernoulli, then, the faster top air exerts less pressure on the wing than the slower bottom air does, so the wing is pushed upward by a net force called *lift*.

That's all very well except for one thing: The top air and the bottom air *don't* have to reach the trailing edge of the wing at the same time; the equal transit time assumption is just plain wrong, in spite of all the arm-waving that physics teachers and flight instructors do to try to justify it. You and I can both forget our embarrassment at never having understood that point in school. There is simply no good reason that the top air has to arrive at the trailing edge at the same time as the bottom air.

The Bernoulli effect does contribute some lift to an airplane wing, but acting by itself it would require a wing that is either shaped like a humpback whale or traveling at an extremely high speed.

Thank you, Mr. Bernoulli. You may step down now.

We now call Sir Isaac Newton to the stand.

Newton's three laws of motion are the ironclad foundation of our understanding of how things move. Newtonian mechanics (as distinguished from quantum mechanics and relativity) can explain the motions of all objects, as long as they are not too small (smaller than an atom) and are not traveling too fast (near the speed of light). Newton figured out his laws for the motions of solid objects, but they can be applied as well to the interactions between airplane wings and air. Let's see how.

Newton's Third Law of Motion (again) says that for every action there must be an equal and opposite reaction. So if the plane's wing is being pushed or lifted up, then by gosh something else is being pushed down. It is. The air. The wing must be whooshing a stream of air downward with a force equal to the lift it is getting. We'll call it *downwash*.

How?

When a fluid such as water or air flows along a curved surface, it tends to cling to the surface more tightly than you might expect. This phenomenon is known as the Coanda Effect. (See the explanation on page 188, but instead of water flowing over a curved glass surface, think of air flowing over a curved airplane wing.) Because of this clinging, the air flowing over the surfaces of the wing is constrained to hug the shapes of the wing; the top-of-the-wing air clings to the top surface and the bottom-of-the-wing air clings to the bottom surface. The streams not only take different paths, but as a consequence of the wing's shape they wind up flowing in different directions at the back of the wing. It's not as if the wing were simply cutting through the air like a flat knife blade, with the airstream parting to let it through and then closing back to its original direction after the wing passes.

As the top-of-the-wing air meets the leading edge of the wing it flows first upward over the surface and then downward again as it leaves the trailing edge. *But the shape of the wing leads it farther downward than where it began; it leaves the trailing edge of the wing in a net downward direction.* In other words, the top-of-the-wing air is actually being thrust downward by the wing's shape. And according to Newton's Third Law, the wing is therefore thrust upward with an equal amount of force. Voilà! Lift!

Do you think this can be only a small amount of force, coming as it does from a push by "thin air"? Hah! Think again. Even a small plane like a Cessna 172 flying at 110 knots (204 kilometers per hour) is pumping three to five tons of air downward every second. Just think of the hundreds of thousands of tons of air that an 800,000-pound (360,000-kilogram) Boeing 747 is pumping downward every second to get off the ground and stay there.

We can give Isaac Newton still more credit for lifting airplanes, because the lift doesn't all come from downwash (with a slight assist by Mr. Bernoulli). Some of it comes from yet another application of Newton's Third Law. Airplane wings are not parallel to the ground; they are made to be tilted slightly upward in front—usually about 4 degrees when the plane is in level flight. That makes more pressure on the bottom surface than on the top, thereby pushing the wing upward and contributing to the lift (see p. 22). The pilot can tilt the plane even farther upward in front (Flyspeak: He can increase his *angle of attack*) to get even more lift from this effect. Sir Isaac's Third comes in because as the plane moves, the wing is pushing the air down in front of it, so the air responds by pushing the wings up.

We see, then, that two different wing actions create lift: the wings' shape—the "airfoil"—and their upward tilt, or angle of attack. Both must be used to maximum effect in order to grunt a heavy plane off the ground during takeoff. That's why you see planes taking off from the airport at such steep angles of climb; the pilots must increase their angle of attack to gain extra lift while the plane is so loaded down with fuel, not to mention that fat lady in the seat next to you.

And you thought the pilot was simply pointing the plane's nose in the direction he wants it to go in, as if it were a horse.

BONUS: Have you ever wondered why ski jumpers bend over so far forward when they're in the air that their noses practically touch the tips of their skis? Two reasons. First, if they stood straight up they'd encounter more air resistance, which would slow them down. But second, their arched backs simulate an airfoil. Their upper surfaces are curved like an airplane wing, and they actually gain some lift that keeps them in the air longer.

Flying with the Top Down

If an airplane's wings are shaped to give it lift, how can an airplane fly upside down?

It can be done to wow the crowd at an air show, but it wouldn't work for a commuter flight to Schenectady because, although it's theoretically possible, passenger planes aren't built to stand the stress. (Nor are the passengers.)

A conventional airplane's wings are curved or humped on top, and that produces lift for reasons that are far from simple (see p. 18). But if the wing were upside down, wouldn't that produce the opposite effect, turning the "*lift*" into "*plunge*"? Yes, if the pilot weren't partially offsetting that effect by changing the plane's *angle of attack*, the angle at which the wings hit the air.

TRY IT   Stick your hand out the window of a speeding car—not above the speed limit, of course. When you hold your hand flat, palm parallel to the ground, you feel the air's pressure on what pilots would call the leading edge of your hand—the thumb edge. But then if you tilt your hand slightly upward, so that your palm gets the brunt of the wind, your wing—uh, hand—is pushed upward. There's more push on the bottom than on the top, and that makes lift, no matter how your hand—or a wing—might happen to be shaped, as long as it's reasonably flat.

So when flying upside down, the stunt pilot points his nose (the plane's, that is) upward, so that the bottoms of the wings—which used to be the tops—are getting the brunt of the wind and are being forced upward. As a matter of fact, stunt planes don't even have wings that are more curved on top; the top and bottom surfaces are the same shape, so it doesn't matter which side is up—everything is accomplished by angle of attack.

As you saw from your hand-out-the-window experiment, increasing your angle of attack produces not only lift, but *drag*—more wind resistance trying to hold your hand back. Similarly, when the stunt pilot increases his angle of attack, the plane experiences more drag against which the engines have to labor. Stunt planes, therefore, have to have powerful engines, as well as crazy pilots. Well, crazy like a fox, perhaps, because it takes great strength and presence of mind to think in three dimensions while you're being subjected to forces that are eight or ten times as strong as gravity. And stunt pilots aren't protected by "g-suits," those pressure suits that fighter pilots wear to keep the blood from leaving their heads and blacking them out during high-acceleration maneuvers.

All the same, I'll just watch from the ground.

How Astronauts Lose Weight

Does gravity peter out at a certain distance from Earth? Otherwise, how can orbiting astronauts be weightless?

Answer to the first question: No.

Answer to the second question: They're not weightless. There's a completely different reason why astronauts can do all those silly tricks for the cameras, such as performing somersaults in midair or sitting upside down on absolutely nothing, looking more witless than weightless.

Earth's gravitational attraction, like all gravitational attraction, reaches out indefinitely; it keeps getting weaker and weaker the farther away you go, but it never diminishes to zero. Every atom in the universe is gravitationally pulling on every other atom, no matter where. But of course, the bigger the agglomeration of atoms you have, such as a planet or a star, the stronger will be their cumulative pull.

That's all beside the point, however, because the paltry 250-mile (400-kilometer) altitude at which the space shuttle goes 'round and 'round is peanuts as far as gravitational weakening is concerned. After all, Earth holds on to the moon pretty well, doesn't it? And that's 239,000 miles (385,000 kilometers) away. (Okay, so the moon is much more massive than an artificial satellite and the strength of the attraction is proportional to the mass, but you get the point.)

If those floating folks aren't weightless, what do we mean by weight, anyway?

Weight is the *strength* of the gravitational pull that Earth exerts on an object. Because that strength diminishes the farther an object goes from the center of the Earth (see pp. 104 and 122), its "weight" diminishes also. But never to zero.

Okay, then. If orbiting astronauts aren't exactly weightless, how come they can float around in the shuttle like that? The answer is that their still-considerable weight is *counteracted* by something else: a force that comes from their orbital speed. (Techspeak: *centripetal force.*)

TRY IT   Tie a string firmly to a rock and swing it around in a circle (outdoors!), holding your hand as stationary as possible. The rock is the shuttle and your hand is the Earth. Why doesn't the rock fly off? Because by means of the string you're pulling on the rock with exactly enough force—an imitation gravitational force—to counteract its tendency to fly away. Pull a little less hard (let some string slip out) and it flies outward, away from your hand. Pull tighter by pulling the string in (imitating a stronger

gravitational attraction) and the rock "falls" inward
toward where your hand used to be.

It's the same with the shuttle. The fact that the shuttle
keeps going around in a stable circle rather than flying off
into space means that its tendency to fly away from Earth is
being exactly counterbalanced by Earth's gravitational
pull, which holds it down. In other words, gravity is continually making the shuttle "fall" toward Earth, exactly
enough to keep it from "rising" farther above Earth. (See
also p. 26.)

The same thing is happening to the astronauts inside.
Their tendency to fly away from Earth is exactly balanced by
Earth's pull, so they neither fly away nor fall to Earth; they
stay suspended in midair, not knowing which way is up.
Which is perfectly okay, because there is no "up." "Up" has

always meant "in the opposite direction from gravity's pull," and gravity's pull is no longer discernible. That's why it's so much fun for them to pose for the camera with one guy upside down. Or is it downside up?

Incidentally, the fact that Earth's gravitational force is balanced by the orbiting astronauts' centripetal force doesn't entirely exempt them from the effects of gravity. It's only *Earth's* gravity that is balanced out. The moon, the planets, the shuttle and the astronauts themselves still attract one another because they all have mass. But because the moon and planets are so far away, and because the astronauts and their equipment don't have much mass, all these gravitational effects don't amount to much. They're still there, however, and that's why space scientists never talk about zero gravity; they say that the astronauts are operating in an environment of *microgravity*.

Up, Up . . . and Around!

How high does a rocket have to go before it can orbit around Earth?

It's not how high—it's how fast. There is a certain speed called the *escape velocity* that an object must achieve before it can keep circling Earth in a stable orbit and not fall down.

Let me take you out to the ball game.

Suppose that a center fielder tries to throw a runner out at home plate with a single mighty throw instead of relaying it via the second baseman. He throws the ball horizontally or slightly higher than horizontally, straight at the catcher. If there were no gravity (and no air resistance), the ball would continue in a straight line and go on forever. Or as Isaac Newton said in his First Law of Motion, "An object will continue moving in a straight line at a constant speed unless some other force screws it up." (He may not have said it exactly that way.)

But in this case there *is* another force: gravity, which is pulling continuously down on the ball whether it is moving or not. The combination of horizontal motion from the throw and vertical motion from gravity results in the ball's following a curved path or trajectory. Unfortunately, few outfielders can throw as fast and far as is necessary to pick off a runner at home plate, so the ball will hit the dirt well in front of the catcher.

Now let's ask Superman to throw a baseball horizontally out over the Pacific Ocean. (And again we'll ignore the air's resistance.) If he throws the ball at, say, 1,000 miles per hour (1,600 kilometers per hour), its curved path will be a lot longer and broader than in the case of the outfielder, but eventually, gravity will still be able to bring it down after perhaps a few miles.

Embarrassed by this pipsqueak performance, our hero then winds up and hurls another baseball out over the ocean at 25,000 miles per hour (40,000 kilometers per hour). This time the ball's trajectory is such a broad, shallow, flat curve that it matches the curvature of Earth's surface itself, so it just keeps going at a constant height above the surface and never falls down. It has gone into orbit.

So you see, putting a baseball or a satellite into orbit is purely a matter of throwing or shooting it fast enough that its trajectory will match the curvature of Earth. That speed, the escape velocity, is 6.96 miles per second (11.2 kilometers per second) or just about 25,000 miles per hour (40,000 kilometers per hour). Any slower than that and gravity will bring the object down before it has gone full circle around Earth. Any faster than that and it will still go into orbit, but it will reach a higher altitude above the surface before gravity wins out and bends its trajectory to the curvature of Earth.

In a very real sense, the orbiting baseball or satellite never does stop trying to fall to the surface; it's just that it is going fast enough "outward" to counteract gravity's inward pull.

That's why physicists and space scientists say that an orbiting satellite or space shuttle is in continuous *free fall*, falling freely toward the center of the Earth (see p. 104), just as if it had been dropped from a height. And that's why the astronauts inside an orbiting shuttle float freely in the air (see p. 23), just as they would if they were in a falling elevator whose cable had snapped (see p. 6).

You Didn't Ask, but . . .

If Earth is spinning, why doesn't the atmosphere go flying off into space?

In order to leave this planet, the air—just like anything else—would have to be moving at a speed equal to the escape velocity. That would amount to a humongous wind. While Earth's motion does affect the winds (see p. 125), the effect is nowhere big enough to get them to blow as fast as the escape velocity.

Individual air molecules may reach escape velocity, however, and some of the lightest atoms such as hydrogen and helium do indeed go into orbit at the top of the atmosphere.

Eavesdropping on the Lake

Sometimes when I'm in my summer cabin on the lakeshore at night, I can hear actual conversations of people on the opposite shore, even though it's half a mile or more away. How come?

It's as if the lake magnifies the sound somehow, isn't it? But it isn't actually magnifying the sound, as a microphone and amplifying system would do; it's just that more of the sound is being funneled toward your ears.

Sound consists of vibrations of the air. The guy on the other side of the lake makes sounds by forcing air from his lungs over his vocal cords, which makes them vibrate. They, in turn, make the air exiting his mouth vibrate. He shapes these vibrations into words with his lips and tongue, and the modified vibrations are transmitted through the air to you as air-pressure waves, similar to ripples moving across the surface of water.

As you can see by dropping a rock into a quiet pool of water, water waves spread out equally in all directions. It's the same with sound waves, but in three dimensions; they spread out through the air in all directions: up, down, north, east, south and west. Naturally, when you're at some distance from the speaker you will be able to hear—that is, your ears will intercept—only a small fraction of the spreading waves. The farther away you are, the smaller the fraction of the total sound energy your ears will be able to intercept, because most of it has gone in other directions and the farther away you are, the more "other directions" there are. At half a mile away, the fraction that reaches your ears is usually so small that you can't hear the guy at all if he's speaking at a normal conversational level.

The unusual effect that you're describing has to do with the fact that sound travels slightly faster in warm air than in cool air. That's because air molecules can transmit vibrations only by actually colliding with one another, and warmer molecules collide more frequently because they're moving faster. So we have to take a close look at the temperature of the air above the lake, to see what temperature effects there might be and how they might affect the sound.

During the day, the sun had been beating down on the air and water. But compared with air, water is very hard to heat up, so the water remained cooler than the air. (You may even have jumped into the lake to cool off, right?) The cool water cools the layer of air immediately above it, so that there is now a layer of cool air beneath the upper layers of warmer air. And if there is no wind to scramble up the air layers, they'll stay that way into the evening.

You, at the edge of the lake, are pretty much in the cool layer. The sound coming from bigmouth across the water travels mostly through the upper, warm layer, but when it gets to you it encounters cooler air and slows down. This sudden slowing down of the sound waves makes them bend downward; they are *refracted,* just as light waves are bent downward when they are slowed down while going from air into water (see p. 61). You can think of it as the faster, upper sound waves overtaking the slower, lower sound waves and tumbling over them, so that the sound spills downward. Thus, an unusual number of sound waves are aimed downward to your ears and you hear more than you have any "right" to hear, based solely upon your distance.

Of course, this works both ways. So when you're sitting on your cabin porch in the early evening hours on a calm, summer day, watch what you say—especially about that jerk on the other side of the lake.

Listen Fast!

If I could drive my car faster than the speed of sound, would I still be able to hear the radio?

As your question implies, this is purely an exercise in "What if?" Automobiles, of course, aren't built sleek enough or strong enough to exceed the speed of sound or to withstand the physical stresses of the sound barrier (see p. 147). But it's fun to think about.

The answer is simple: Yes.

Or, I could have posed a different question that would settle the issue: On the supersonic Concorde airliner, can the passengers converse? At those prices they'd *better* be able to. But how, if they are traveling faster than sound?

Even if you were driving faster than the speed of sound, you and the car and the radio and your terrified passengers would all be moving at exactly the same speed relative to the countryside. You're all in the same boat, so to speak. As far as sound is concerned, the important thing to realize is that you and the radio and the air in between aren't moving *relative to one another.* The radio has the same spatial relationship to you as if the car were standing still. It emits its sound waves through the car's air to your ears with the speed of sound as if nothing unusual were happening, because inside the car, nothing is. In fact, if the speedometer and windows were blacked out (God help you), you wouldn't even know you were moving except for the noise and vibration from the wind and the tires (see p. 222).

What if you were driving a supersonic convertible with no windshield and the radio speaker in the back? Could you still hear it? No. Not even considering the effects of the wind on your poor, battered ears and brain, you wouldn't be able to hear the radio. The sound waves from the speaker are being transmitted through the air toward you at the speed of sound, but the air itself—the transmission medium for the sound—is moving backward away from you even faster. So the sound will never reach you. The sound is like a rowboat rowing upstream more slowly than the water is flowing downstream.

By the way, the radio receives its signals by radio waves, not sound waves, and radio waves travel at the speed of light, which is a million times faster than the speed of sound. So any motion of your car is certainly not going to have any effect on the radio's ability to play.

Now what about the sounds that escape from your car? What would a roadside cow hear? (You're not doing this on city streets, I hope.)

Your car noises, whether from radio, tires, engine or screaming passengers, are being sent out in all directions at the speed of sound. But you are approaching the cow faster than that; you are actually outrunning your own sound. As your car approaches the cow, then, she can hear none of the car noises that are trailing behind you until shortly after you pass, when she will hear a sonic boom (see p. 147) and all the car noise.

Note that if you are outrunning sound, you won't be able to hear any sounds coming from behind you, because they can't catch up with you. That's why you can see the flashing lights on that police car that's chasing you, but you can't hear the siren. I doubt, however, that the trooper will accept that as an excuse.

2

Looky Here!

. . . And God said, "Let there be ultraviolet, visible and infrared radiation."

Well, that's not an exact quote, but it certainly was a good decision. The Lord's Lightbulb, the sun, is the source of not only light, but all the energy we use on Earth (see p. 133), with the exception of energy from nuclear reactors, which humans invented in 1942, and Earth's own deep-down heat energy, which we are only now beginning to tap for practical purposes.

But the most apparent role that Old Sol plays—the only one, in fact, that most people ever think about—is that it provides the light we see by, the purifying light of day that brightens and illuminates all of Earth.

When any light—solar or artificial—strikes an object, some of it bounces off (is reflected), some of it is absorbed and transformed into heat, and some of it may even go straight through, as in the case—fortunately—of air, water and glass.

This chapter is a biography of light—what it is made of, where it comes from and goes to at its incredible speed of 186,000 miles per second (3 million kilometers per second), and how it can entertain us, trick us and burn us. As we follow this path of enlightenment we'll have occasion to play in the snow, go to the movies, watch television with a magnifying glass, cool ourselves with an electric fan, fool around with mirrors and even eat some candy that makes sparks in the dark.

Brighter Than Bright

Those brilliant Day-Glo colors—they're unreal! How can they be so much brighter than anything else? They look as if they're actually generating their own light.

They are.

In a Day-Glo-colored object there's a chemical that takes invisible ultraviolet radiation out of the daylight and converts it into visible light of the same color as the object. Thus, the object is not only reflecting its normal amount of colored light, but is actively emitting some light of the same color, which makes it look "extra-colored" and up to four times brighter.

The Day-Glo Color Corporation of Cleveland is only one manufacturer of what are called daylight fluorescent pigments. As the self-proclaimed world's largest supplier, it makes a dozen different colors, from aurora pink to saturn yellow. It sells the pigments to companies that put them and similar dyes into everything from orange safety vests and traffic cones to yellow tennis and golf balls and highlighting pens.

What's going on is *fluorescence,* a natural process by which certain kinds of molecules absorb radiation of one energy and re-emit it as radiation of a lower energy. The molecules in the pigment are absorbing ultraviolet radiation, a kind of short-wavelength radiation that human eyes can't see, and re-emitting it as a longer-wavelength light that human eyes can see. The radiation is, in effect, shifted from invisible to visible.

How do molecules absorb and re-emit radiation?

Molecules contain lots of electrons that have certain specific amounts of energy characteristic of the particular molecule. But these electrons are always willing to take on certain amounts of extra energy from outside. (For more on this point, meet me in the Nitpicker's Corner.)

A molecule of a typical pigment may contain hundreds of

swirling electrons of various energies. When a bullet of ultraviolet radiation (Techspeak: a *photon;* see p. 170) hits such a molecule, it may kick some of those electrons up to higher energies. (Techspeak: The electrons become *excited;* honest, that's what scientists say.) But they can hold on to their overabundance of energy for only a few billionths of a second (a few nanoseconds) before spitting it back out as radiation again—usually as several photons of lower energies or longer wavelengths. It's sort of like spitting out buckshot after stopping a bullet.

Now the "buckshot" radiation, having somewhat less energy than ultraviolet radiation, falls into the region of radiations that human eyes can see: colored light. The net result is that the pigment molecule has absorbed invisible radiation and re-emitted it as visible radiation.

As long as the pigment molecules are being exposed to ultraviolet radiation—and daylight contains lots of it—they will be absorbing it and emitting light of a visible color. If the pigment happens to be orange to begin with and the emitted light is also orange, the dyed object will be an unnatural super-orange—"oranger" than you think it has any right to be.

TRY IT   Shine an ultraviolet lamp—a so-called black light lamp—on a Day-Glo fluorescent object, such as a paper with a few streaks of fluorescent highlighter on it or, if you're the type who wears them, your Day-Glo-printed T-shirt. The fluorescent dye will glow very much brighter than it does in daylight because the lamp puts out much more ultraviolet radiation. If you don't want to buy an ultraviolet lightbulb, take your streaked paper or T-shirt into one of those tacky stores that sell tasteless gifts and fluorescent posters, and use their black light for free.

By the way, if you use a fluorescent yellow highlighter on your books or notes, remember that it is brightest in daylight, which

contains plenty of ultraviolet. Ordinary household incandescent lightbulbs give off very little ultraviolet light; moreover, their light is somewhat yellowish, and that washes out the yellow highlighter color. So when reviewing your highlighted book passages or notes by lamplight the night before an exam or a presentation, you may find to your chagrin that your highlighting is all but invisible. It's safer to use the stronger highlighter colors: orange, green or blue, whether fluorescent or not.

In my work as a professor, just about the only excuse I haven't heard from a student who did poorly on an exam is that the highlighting on his notes disappeared.

You Didn't Ask, but . . .

Why does a white shirt glow brightly under "black light"?

It's the same fluorescence phenomenon as the Day-Glo colors. Most laundry detergents contain "brighteners" (see p. 194) that absorb ultraviolet radiation from daylight and re-emit the energy as a bluish light that makes the shirt look "whiter and brighter."

Moreover, the blue cancels out any yellowish cast. When stimulated by an ultraviolet lamp, which is even richer in ultraviolet radiation than daylight is, the fluorescence becomes bright enough to see as an actual glow-in-the-dark luminescence.

You Didn't Ask This Either, but . . .

How do those luminous light sticks work?

You mean those plastic rods full of liquid chemicals that are made by Omniglow and other companies and are sold at street fairs, festivals and concerts and that start glowing with green, yellow or blue light when you bend them, and that gradually lose their light after an hour or so? Never heard of them.

Okay, seriously.

By now you know that a fluorescent dye needs to be stimulated by absorbing energy before it can re-emit that energy as light. But the stimulating energy need not be visible light or ultraviolet radiation; it can also be heat, electrical or chemical energy. In the case of the light sticks, the stimulating energy is chemical. When you bend the stick, you break a thin glass capsule containing a chemical, usually hydrogen peroxide, that reacts with another chemical in the tube. The reaction gives off energy, which is taken up by a fluorescent dye and re-emitted as light. As the chemical reaction gradually plays itself out because the chemicals are used up, the light fades.

NITPICKER'S CORNER

At several places in this book I talk about a substance's absorbing certain colors or wavelengths of light. You may be wondering how molecules actually absorb light, and what determines which wavelengths they absorb. If that problem is not exactly keeping you awake at night, Nitpicker's Corners are designed to be skippable.

A molecule has custody of all the electrons that belong to the atoms that it is made of. (Molecules are nothing but atoms glued together.) But electrons, and for that matter all subatomic particles, have a peculiar property: They can have only certain amounts of energy and no others. (Techspeak: The electrons' energies are *quantized*.) For example, the electrons in a certain kind of molecule can have energies A, B, C or D, etc., but never A-and-a-half or C-and-two-thirds. They can change their energies up or down among the values A, B, C, D—that is, from A to B or from D to C and so on—but they can never have values anywhere in between. Nobody can give you a reason for this; that's just the way it is. When you get down to things smaller than an atom, it's a different world from the one we see every day up here in big-land.

Now inasmuch as each unique substance is made up of its own unique molecules, it will have its own unique collection of electrons with their own unique sets of allowable energies. When light energy falls upon the substance, its electrons will absorb only those energies that correspond to its allowed energy jumps from A up to B or C, etc; it will reject and reflect the rest. This means that *the substance is actually picking out the light energies (wavelengths) that it prefers,* leaving the others to bounce back as reflected light.

And that's why every colored substance has its own color: the color of those wavelengths that it cannot absorb and that it reflects back for our eyes to see.

Snow White and the Seven Hues

Why is snow white? It's made of water, and water is colorless. So how come it turns white just by freezing?

First, we have to look at what "white" is.

You've heard people say dozens of times that white light is the presence of all colors. But other people tell you that white isn't a color at all, that it's the absence of color. You use bleach to remove all color from your laundry and make it white, don't you? So how can white be both all colors and no color?

The answer is that these two groups of well-meaning people are talking about two different things: white light and white objects.

White light, as it comes to us from the sun, is indeed a mixture of all possible colors—all visible wavelengths. Because we "grew up" as a species with sunlight as our natural, neutral, everyday light, we named it "white," a word that has its origins in an Indo-European word meaning "bright" or "gleaming," with no color implications at all. So white light is colorless light—to human eyes.

But in 1666 Sir Isaac Newton discovered that this neutral light can be broken down into a rainbow of component colors, simply by passing it through a triangular chunk of glass—a triangular prism. He then proved to himself that all these colors were indeed present in the original white light by recombining them: He projected overlapping rainbows onto a wall and saw that they combined to form white light.

Newton thought it would be a nice idea to divide the entire rainbow or spectrum of colors (and he invented the word "spectrum" for the purpose) into seven categories that would be analogous to the seven musical tones in an octave. For his color categories, he chose red, orange, yellow, green, blue, indigo and violet. Unfortunately, more than three centuries later we are still being taught in school that these are "the seven colors of the rainbow"—even though nobody seems to know what "indigo" is. Sir Isaac had to fudge a bit to eke out seven color names.

In reality, there are an infinite number of colors—both visible and invisible to human eyes—in the sun's light, just as there are an infinite number of possible musical tones. Change the wavelength of light or sound by an infinitesimal amount and you've got a brand-new color or tone, irrespective of whether humans can detect the difference. For example, there are dozens of different hues that we lump under the term "red," limited only by our eyes' ability to distinguish them. It is said that the human eye can distinguish as many as 350,000 different hues. (Whose eye? I wonder.)

A white *object,* as distinguished from white *light,* is white because when white light falls upon it, it reflects all those zillions of colors back to our eyes equally, without changing the composition of the mixture at all. Its molecules just don't happen to be absorbers of visible light, so it appears to be the same "color" as the light that fell upon it: what we choose to call "white." The object contributes no color of its own.

But colored objects are indeed contributing colors of their own. Their molecules are selectively absorbing and retaining certain of the sunlight's colors (see p. 34), reflecting back the others as an altered mixture.

Think of an actor on the stage, wearing a red cape over a white shirt. If you shine a red spotlight on him, he will appear red all over, the shirt as well as the cape. That's because the only light that any part of his costume can reflect back to us is red. No part of him can reflect back any green or blue light because it simply isn't receiving any.

Now shine a white spotlight on him. The red cape is still red, because that's the nature of the dye in it; that particular chemical was chosen because it absorbs all other colors in the white spotlight, reflecting back only the red. But the white shirt doesn't absorb *any* of the colors in the spotlight; it has no red dye in it (until the actor surreptitiously applies some in the stabbing scene). The shirt just sends the whole mixture of spotlight colors back to us, looking just as white as when they came out of the can.

Now let's get back to the snow before it all melts.

Snow is white, you now know, because its molecules reflect back to us all the colors in the sunlight. It doesn't selectively absorb any particular colors.

"But wait a minute," you're thinking, "neither does *liquid* water; it's made of the same H_2O molecules. So why isn't liquid water as white as the driven snow?"

Because liquid water is a poor reflector. When light hits it squarely, almost all of it goes straight in—penetrates—rather than bouncing back. In other words, liquid water is transparent. And if practically no light bounces back, it can't display much color, not even whiteness.

Snow, on the other hand, is an excellent reflector of light—whatever kind of light hits it. You want green snow? Hey, Sammy! Turn on the green lights! It's an excellent reflector because, unlike liquid water, which passively allows light to penetrate it, snow consists of zillions of ice crystals, each one a tiny jewel with dozens of sparkling facets that reflect light like mirrors. All of this white light bouncing back to our eyes, with its full complement of original colors intact, is what makes snow appear even whiter than the actor's sweaty shirt.

You Didn't Ask, but . . .

What is black? Is it a color?

A black surface is one whose molecules are absorbing all visible wavelengths of the light that is falling upon it, and reflecting virtually none of it back. So black isn't really a color, because we define a color in terms of the specific combination of light wavelengths that reflect back into our eyes.

But, of course, you can *see* a black object, so it must be reflecting *some* light back to our eyes. Hey, who's perfect? The light that a black object is reflecting comes from the fact that its surface has a small but unavoidable amount of shininess to it. So it reflects back some of the light that hits it at a glancing angle. That's why there are "light black" and "dark black" objects, depending on the microscopic glossiness of their surfaces. Go to a hardware store and look at the black paints; they're all equally black, but they'll range in reflectivity all the way from flat to glossy.

NITPICKER'S CORNER

I said that when light hits liquid water squarely, most of it goes straight in without being reflected. The emphasis was on that word "squarely." As I'm sure you've observed, the surface of water can be a very good reflector of light that is hitting it obliquely—at a glancing angle. When the sun is low over a lake, for example, its reflection on the water can be almost blinding.

It's the same with snowflakes. Yes, they're made of transparent crystals of ice, but because there are zillions of them, all with complicated shapes, scattered helter-skelter with their smooth reflecting facets facing in all directions, the light is almost invariably striking the facets obliquely and being reflected. That's why snow is such a good reflector—so good, in fact, that skiers and other masochists who like to frolic in frigid weather have to wear very dark glasses to avoid "snow blindness."

A final point. I let you get away with thinking that liquid water is colorless. It's *almost* colorless, but not quite. (See p. 179.)

School Colors

In science class they told us that the primary colors are blue, green and red. But in art class they told us that the primary colors are blue, <u>yellow</u> and red. Why can't artists and scientists agree?

Because they think of color differently.

Scientists describe objectively what Nature provides. They therefore think of color as a fundamental characteristic of light itself. To a scientist, light of different colors is radiation of different wavelengths. Artists, on the other hand, create their own interpretations of Nature. They therefore tend to think of color subjectively, as something to be manipulated with paints and dyes, rather than accepting light in its natural state.

Why, then, do these two camps have to use different primary colors—trios of colors that can be combined in different amounts to produce all other colors? In a nutshell, it's a matter of the primary colors of *light* versus the primary colors of *pigments*. As we'll see, they can be called the *additive primaries* and the *subtractive primaries,* respectively.

Light-minded (not light-headed) scientists claim that they can make light of any perceived color by combining blue, green and red light of various intensities. On the other hand, pigment-minded artists claim that they can tint an object any color by combining blue, yellow and red pigments in various amounts. And they're both right, because there is a fundamental difference between the color of light and the color of an object.

Colored *light* is a certain color because it is made up of a mixture of light waves of various wavelengths. The different-colored components *add together* to produce the net color. It happens that, because of the way our eyes work, blue, green and red light contain all the necessary wavelengths that need to be mixed in order to produce any perceived color. So blue, green and red are called the primary colors of light. (Understood: for human eyes.)

A colored *object,* on the other hand, is a certain color because of the wavelengths that it *absorbs* from the light that is falling on it (see pp. 37–38). In other words, it *subtracts* certain wavelengths from the light and reflects the rest back to us as the color that we see. Various mixtures of blue, yellow and red pigments are capable of absorbing almost any combination of wavelengths. So blue, yellow and red are considered the primary colors for mixing paints and dyes. (But see below for a little hedging about these three colors.)

The light-based system of primary colors is called *additive* because different combinations of wavelengths add together to produce different colors of light. The pigment-based system of primary colors is called *subtractive* because different combinations of wavelengths are absorbed or removed from light to produce different colors of paints and dyes.

Let's look at the *light* primaries first, then the primaries for *objects* or pigments.

Light: The human eye—even an artist's eye—works on the additive principle. It has three kinds of color-sensitive cells (so-called cone cells) on the retina: One is most sensitive to blue light, one to green and one to red. Our perception of various colors depends on the relative degrees of stimulation of these three types of cells by the incoming light; the brain adds them together to produce sensations of various colors. That's why scientists—human chauvinists that they are—chose blue, green and red as the primary colors of light. (Muskrat scientists undoubtedly use a different set of primary colors.) Our eyes react only to stimulations of those three color receptors, so they are all we need in order to produce all humanly discernible hues. And that's why there are three, and only three, "primary" or fundamental colors of light.

Note that each kind of cone cell is not sensitive *exclusively* to pure blue, green or red light; each one is sensitive to a lesser degree to the other colors as well. That's why we can see pure yellow light, even though we don't have any yellow-sensitive cone cells. The yellow light slightly stimulates both the green and red cells, and our brains perceive that combination as yellow.

Your color TV and computer screens take advantage of the three-color idiosyncrasy of human vision. They contain blue, green and red phosphors (chemicals that glow when stimulated by electrons), glowing with varying brightnesses. The glows all add together to produce the colors that we perceive.

TRY IT   Look at the picture on your color TV or computer screen with a magnifying glass. You'll see that it's made up of tiny blue, green and red rectangles—

no other colors—that are being stimulated to glow with varying brightness. Your eye blends them all together because the individual rectangles are too small to see at normal viewing distance. Added together in this way, the primary-colored rectangles make the hundreds of different hues that you perceive.

Pigments: The color film in your camera, on the other hand, makes its colors by the artist's subtractive system. It contains three layers of dyes that *absorb* or filter out blue, green and red. And the absorbers or filters that best absorb blue, green and red happen to be yellow, red and blue, respectively. So yellow, red and blue are the three subtractive colors in color film.

But are these color-film filters the same old yellow, red and blue colors that your art teacher told you are the artist's subtractive primary colors? Sort of, but not exactly.

Here's the hedging that I promised you: The three colors that are *really* best at absorbing the blue, green and red that

our cones are most sensitive to are yellow, a purplish red called magenta and a greenish blue called cyan. Yellow, magenta and cyan are therefore the three *real* primary subtractive colors that are used in concocting the entire spectrum of ink, photography and paint colors.

All artists, from kindergarten kids with crayons to the subtlest watercolorists, could create their entire palettes by mixing various amounts of yellow, magenta and cyan. But it's a lot easier to buy paints and crayons already blended.

Let There Be Fluorescence!

How do fluorescent lamps make so much light without a lot of heat? And when one burns out, can I replace it with any tube that fits, or are there different kinds?

Fluorescent lights were invented for one purpose: to confuse you. I'm glad to see that they're doing their job.

When an ordinary incandescent lightbulb burns out you can just screw in a new one with the help of a certain number of friends, depending on your vocation or ethnicity. But when a fluorescent light burns out you look at the tube to find out what kind to replace it with and you see markings that look something like "F20CW-T12." If you replace it with the "F15W-A10" that you saw in the store, will it explode when you turn it on?

Cheer up. 'Tis better to light a candle and read this book than to curse the darkness.

First, let's decipher those hieroglyphics on the tube. They're a secret code that divulges everything about the bulb. Not to you, the poor consumer, of course, but to the people who make and sell them, who apparently have a need to appear smarter than you are.

I'm going to tell you how the secret code works. (I suppose that now they'll have to kill me.)

Any given fluorescent tube is either straight, U-shaped or

circular in shape; it has a certain wattage; it gives off a certain color of light; and it has a certain diameter. The letters and numbers on the tube give this information in that order: shape, wattage, color, diameter. The only trouble is that you have to know how this information is coded.

For shape, it's a U or a C for U-shaped or circular, and no letter at all if it's straight. Then comes the wattage: 4, 5, 8, 13, 15, 20, 30, 40 or whatever. (The wattage is generally lower than for comparable light-producing incandescents, because fluorescent lighting is from two to four times more efficient.) Then comes the color code: W for white, CW for cool white, WW for warm white, plus abbreviations for other exotic colors that we needn't bother with. Last comes the tube's diameter, but it is given—would you believe?—in eighths of an inch: T8 means a tube that is eight-eighths of an inch in diameter, which any sane human being would call one inch. A T12 tube is twelve-eighths or one-and-a-half inches in diameter, and so on.

Pop Quiz: Describe the properties of an F40CWT10 fluorescent bulb. (Answer at the end of this section.)

Oh, I forgot to tell you: The codes always begin with an F for "fluorescent," presumably to keep you from trying to screw them into an ordinary lamp socket. (How many idiots does it take to screw a fluorescent tube into an incandescent lamp socket?)

As an alert consumer, you may have noticed that you can't replace an 18-inch-long tube with one that is 24 inches long. The manufacturers graciously give you enough credit to make that decision on your own, so you won't find a length code on the tubes.

Okay, now. How do the things work?

You know that ordinary incandescent lamps, including halogen lamps (see p. 50), make light by electrically heating a filament to white heat. The outside of the lamp bulb can get up to temperatures of several hundred degrees. Fluorescent lamps work on an entirely different principle.

The fluorescent tube is filled with a small amount of inert gas (usually argon) plus a drop's worth of mercury. At each end is a small filament that is heated by the electric current so that it emits electrons. (You don't know why a hot filament emits electrons? Go stand in the corner. The Nitpicker's Corner, that is.)

The electrons emitted from the filaments fly through the gas in the tube to get from one filament to the other, and in the process they collide with mercury atoms, which have been vaporized by the filaments' heat. The mercury atoms absorb the collision energy and spit it out again as light energy. But we can't see that light because it's in the ultraviolet region of wavelengths, so it has to be converted into light that humans can see. This is accomplished by that white coating on the inside of the tube. It consists of chemicals (calcium and strontium phosphates and silicates) that absorb ultraviolet light and re-emit it as visible light; this wavelength-shifting process is called *fluorescence* (see pp. 34 and 37).

Fluorescent lamps are cooler than incandescent lamps because they have only those two little mildly heated filaments at the ends and the fluorescence process itself doesn't produce any heat. But they're hard to start, because the filaments' electrons first have to blast their way through the gas in the entire length of the tube. That requires several hundred volts of push (see p. 100), but our household voltage is only 115 volts. So something has to provide an initial voltage kick to the electrons.

That's what the starter does—or the ballast. And here's where it gets really confusing, because there are several kinds of fluorescent lamp systems and circuits. Some have ballasts, those heavy little iron transformers, while others have starters, those little aluminum cans. And some have both. Fuhgeddaboudit. You don't hafta know.

What to do when your fixture of unknown breed won't light up? First, replace the tube with one that has identical code numbers on it. You can't even substitute a different wattage, as you can with incandescent bulbs; that can cause dangerous overheating of the ballast, which was designed for the other wattage. The only freedom you have is to swap a cool white for a warm white or vice versa, or to substitute one of the many other "deluxe" colors. If your fixture has one of those little starter cans in it, you may as well replace that too; they're cheap and they simply twist in and out of the socket.

If you're still in the dark, both literally and figuratively, buy a whole new fixture.

Oh, and an F40CWT10 is a 40-watt, cool white, one-and-a-quarter-inch straight fluorescent tube.

You Didn't Ask, but . . .

Why do small fluorescent tubes cost so much more than the big four-foot-long "shop lights"?

You can buy the common, 48-inch shop-light tube in home centers for a couple of dollars, whereas a small, thin fluorescent tube for under the kitchen cabinet might cost up to five times that amount. The answer is that the four-footers, used by the thousands in schools, factories and office buildings, vastly outsell the smaller, more specialized tubes and are mass-produced at a much lower unit cost. It's a classic textbook case of supply and demand.

NITPICKER'S CORNER

Why do the filaments in a fluorescent tube emit electrons when they're heated?

Almost anything will emit electrons if heated hot enough. Atoms contain negatively charged electrons, which are held on to with various degrees of strength, depending on which atoms we're talking about. Metal atoms hold on to their electrons very loosely. When you heat a metal, some of those electrons gain enough energy to detach themselves completely from their atoms and go flying off.

In a fluorescent tube there are two filaments, one at each

end, both getting hot because of their resistance to the flow of a sixty-cycle alternating electric current (a current that continually reverses its direction). At any given instant, one filament is negatively charged with respect to the other, but a hundred-twentieth (half a sixtieth) of a second later it becomes positively charged with respect to the other. At any instant, the electrons from the negative filament are attracted to the positive filament, and the only way they can get there is to plow through the intervening mercury vapor in the tube, making it emit ultraviolet radiation.

Star Light, Star Bright, Which Bulb Should I Use Tonight?

What's so special about halogen lightbulbs?

They contain a gas called a halogen, which makes them brighter, whiter, more efficient and longer-lasting. And, of course, much more expensive.

A halogen lamp is a variation on the standard *incandescent,* as opposed to *fluorescent,* lamp (see p. 46). An incandescent lamp contains a tungsten filament enclosed in a glass bulb filled with gas. An electric current heats the filament to incandescence—a white-hot glow. It may look very bright, but in reality only 10 to 12 percent of the energy it emits is visible light; about 70 percent of it is invisible infrared radiation (see p. 112), which heats, rather than illuminates.

In a regular bulb, the gas inside is an inert (unreactive) one such as argon or krypton with some added nitrogen. These inert gases keep the tungsten from oxidizing, or "burning up," as it would in air. Some smaller bulbs solve the problem by being completely evacuated; there's practically no gas inside at all.

In a halogen bulb, the gas is usually iodine or occasionally bromine, two highly reactive chemical elements in the family that chemists call halogens. They perform a two-step chemical dance that makes the filament last twice as long. But first, we have to understand how the standard bulb works.

The filament is a coil of thin tungsten wire. Tungsten is used because it has the highest melting point of all metals—6200 degrees Fahrenheit (3400 degrees Celsius)—and it stays strong even at white-hot temperatures of 4500 degrees Fahrenheit (2500 degrees Celsius) or higher. Moreover, it has the lowest vapor pressure (see pp. 208–209) of all metals, meaning that it evaporates less than any others. Yes, even metals evaporate a few atoms now and then, but so slowly that we never notice it except at very high temperatures. (Never fear; your gold jewelry isn't going to dry up.)

When it is white-hot, even tungsten will evaporate enough so that the filament gets thinner and thinner as the bulb burns, until it finally breaks apart and interrupts the electric circuit. That's when your bulb burns out. For some time before this disaster strikes, you can see the evaporated tungsten as a dark coating on the inside of the glass, where it has condensed because of the glass's relatively low temperature. This darkening, of course, progressively cuts down on the amount of light that the bulb puts out as it ages.

Sometimes a bulb's filament will have developed such a thin spot that it will blow out suddenly when you turn on the switch. The blue flash that you see is an electric arc, leaping across the widening gap as the thin spot evaporates completely under the heat stress of the power surge.

Tip: When a bulb burns out, try tapping or shaking it gently while the power is on. Sometimes you can get the broken ends close enough together so that an arc will flow between them and weld them back together, rewarding you with perhaps an hour or so of life-after-death experience.

What halogen-filled bulbs do is to cut down the evaporation rate of the tungsten in a very interesting way. First, the iodine vapor reacts with the evaporated tungsten atoms before they can condense out on the glass and converts them to tungsten iodide, a gaseous chemical compound. The molecules of tungsten iodide then float around inside the bulb until they happen to encounter the white-hot filament, whereupon the high temperature breaks them back down again into iodine vapor and metallic tungsten, which deposits itself back on the filament. The released iodine is then free to apprehend and deliver more tungsten atoms, and the cycle continues, with the iodine atoms continually capturing evaporated tungsten atoms and returning them to the filament. This recycling process approximately doubles the life of the filament, and hence of the bulb.

The halogen process allows the lamp to be operated at a much higher temperature without excessive deterioration of the filament, and that makes a brighter, whiter light. In fact, the temperature of the bulb's inside wall *has* to be high— above about 480 degrees Fahrenheit (250 degrees Celsius)— to keep the tungsten atoms from condensing on it before the iodine vapor can grab them.

Halogen bulbs are made of quartz, which withstands much higher temperatures—and is more expensive—than ordinary glass. They are usually tube-shaped and closely surround the filaments to stay hot. In fact, tungsten lamps burn so hot that they can be a fire hazard if used too close to flammable materials such as curtains.

You Didn't Ask, but . . .

Why don't lightbulbs last longer than they do?

Lightbulbs are very carefully engineered to last for a certain length of time. A suspicious person might be tempted to say that they are carefully engineered to *burn out* after a certain length of time. There is no reason that a lightbulb couldn't be designed to last almost indefinitely. But you probably wouldn't like it.

As with most devices, there is a trade-off among several conflicting considerations. More than anything, the life of a bulb depends on the running temperature of the filament. For a given wattage (the amount of electric power consumption), the higher the temperature and light output, the shorter the lifetime.

"Long-life" lightbulbs have filaments that are designed to glow at a lower temperature. But the lower temperature doesn't produce as much light. Also, since higher temperatures produce a bluer, whiter light, the long-life bulbs can have a slightly yellowish cast by comparison.

Long-life bulbs achieve their lower temperatures by using a filament that allows less electrical current to pass through. Less current flow makes less heat and less light, so you get not only a yellower light, but less of it. If you buy long-life bulbs, you have to buy a higher wattage than usual to get the amount of light you expect from a normal bulb.

By law, the packaging of standard lightbulbs must tell you the number of hours they are intended to last and the amount of light they put out in all directions: the number of lumens. Compare the numbers of hours and lumens on a long-life package with the numbers on a regular package of comparable wattage. If you're willing to put up with the lesser amount of light and higher price for the convenience of not having to change the bulb for a longer period of time, buy the long-lifer.

On the other hand, if you're a compulsive discount shopper for standard lightbulbs, take your calculator to the store. For a given wattage, you want the most light for the longest time at the lowest price. Divide the price in cents by the number of lumens, and then divide the result by the number of hours of expected lifetime. The smallest number is the best bargain.

And speaking of saving money, a dimmer switch reduces the voltage (see p. 97) applied to the bulb, which reduces the current flowing through the filament, which reduces the temperature, which reduces the evaporation of tungsten, which considerably increases the lifetime of the bulb. The next best thing to turning out the lights when you leave a room is to dim them.

Mirror, Mirror, on the Wall, How Come You Don't Invert at All?

When I look in the mirror and raise my right hand, my image raises its left hand. And yet both our heads are still on top. Why does a mirror reverse things right to left, but not top to bottom?

This is one of those loaded questions that can drive you crazy because the question itself is misleading. It starts with a mistaken assertion and asks us to carry on our reasoning from that point. But you can't pursue the road to truth if somebody starts you off in the wrong direction.

A mirror does not reverse things right to left. It reverses things front to back; it reverses in and out.

Read that again.

And think about it.

All a mirror can do is reverse a *direction*. It can't rotate anything. It's *you* that imagines yourself rotated. The mirror didn't do it.

Stand in front of a full-length mirror. Let's name the person in the mirror Egami. Now how do you think Egami got that way, with his left arm toward your right and his right arm toward your left? I'll bet you seven years of bad luck that you think Egami got that way by your turning around—by your rotating half a turn, executing an about-face. That's why you think right and left have been reversed. *You* did it yourself, by turning yourself around—in your imagination.

But that's not what the mirror did.

All the mirror did by taking its incoming light and shooting it back at you was to reverse the direction of the light. Egami is simply you with your "toward" and "away" directions reversed. You are, of course, in the habit of looking *away* from yourself, but Egami is looking *toward* you; if you're facing north, Egami is facing south. And whenever a person is facing in the opposite direction from you and looking toward you, his left arm will be on your right, no? What's so unusual about that? No rotations or right-to-left swaps are needed.

Notice that the words "up," "down," "top" and "bottom" appear nowhere in the foregoing. They're completely irrelevant to two people who are facing each other. "Up" and "down" mean exactly the same thing to both of them. Unless, of course, one of them is standing on his head.

How can we get one of them to stand on his head? Easy. Hold the mirror high above your head and parallel to the floor. Or else put the mirror on the floor and stand near (not on!) it. Egami is now standing on his head, isn't he? Which proves that the mirror reverses only its in and out directions, which from your current viewpoint just happens to be up and down.

You can see the same up-down reversal in the mirrorlike surface of a small, calm lake or pond. Look at the reflection of the trees on the other side. They're upside down, aren't they?

And by the way, I've referred to Egami with masculine pronouns to avoid "him-or-her"-ing all over the place in an explanation that you may think is already complex enough. If you're female, please don't think I'm saying that mirrors reverse gender. (And don't wear a skirt when you put that mirror on the floor.)

Oh, the name? If you haven't yet figured it out, Egami is "image," reversed from right to left.

You Didn't Ask, but . . .

When I look into the bowl of a shiny spoon, my image is both reversed right to left and inverted upside down. How does it do that?

I just finished explaining that the image isn't really reversed right to left, so let's put that aside. But indeed, how about the upside-down inversion?

The spoon's inner surface is concave—that is, it is hollow like a cave. (That's a good way to remember the distinction between con*cave* and con*vex*.) When you look into the spoon, you'll notice that the top part is shaped so that it reflects its light slightly downward, like a mirror held high. At the same time, the bottom part is shaped so that it reflects its light slightly upward, like a mirror on the floor. These "high" and "low" reflectors give you a stand-on-the-head image, exactly as the above-your-head and on-the-floor mirrors did in the preceding explanation.

Mirror, Mirror, on the Wall, Who's the Sharpest One of All?

I'm nearsighted. When I look in the bathroom mirror without my glasses on I can see my beautiful face quite clearly, but everything else in the room is blurred. Shouldn't everything be equally clear, because all the images in the mirror are equally close to my eyes?

The distance from your eyes to the mirror is irrelevant. It's the distance from your eyes to a given object that counts, just as it would when you look at it without the mirror.

The light reflected from an object has to get to your eyes somehow, or else you wouldn't see it. The light coming from things behind your back would never get to your eyes if the mirror weren't there to turn it around. That's all the mirror does: It takes light that would have passed you by and shoots it back at your eyes (see p. 54).

Suppose you're facing the mirror and looking at an object behind you. Instead of coming straight from the object to your eyes, the light has to pass you by, go to the mirror and then come back to your eyes. That's a greater distance than if you had been facing the object, so it is even blurrier than if you had turned around and looked at it directly. The image of your beautiful face is also blurrier than if you were looking at it from the position of the mirror. The light has to go from your beautiful face to the mirror and back to your beautiful eyes—twice as far as if you were looking at your beautiful face from the position of the mirror.

This is all based on the fact that the farther away an object is, the fuzzier it will appear to nearsighted eyes. That's generally true, and here's why.

Nearsighted eyes are good at focusing light rays that are diverging, radiating out in all directions, as they are from a nearby object. But nearsighted eyes are not so good at focusing light rays that are more or less parallel, as they are from a distant object. It's not that near and distant objects are shooting their light out differently; every object reflects light in many directions. (Remember how we drew a shining sun in kindergarten, with all those rays coming out in all directions?) But when you're far away from an object, your eyes are intercepting only a small fraction of those "all directions" rays. It's as if all the rays are now coming from the same, severely limited direction, like a bundle of parallel sticks, all pointing from the object straight at you. And that's the situation—focusing parallel rays—that nearsighted eyes can't handle well, so the object is blurred.

They Went . . . Which-a-way?

In western movies, why do the stagecoach wheels sometimes turn backward?

This is the only remaining artificiality in today's remarkable, computer-driven movie effects, which can make anything imaginable look real, no matter how bizarre—except, ironically, an old-fashioned stagecoach wheel. You can also see the effect with automobile wheels, in those television commercials that show the cars speeding along an open road.

If you watch carefully, you'll see that the wheels go backward only some of the time; at other times they look as if they're rolling forward rather slowly, and at still other times they seem to stop entirely, making the coach look like a sleigh. It's all a matter of timing—the speed of the rolling wheel compared with the speed of the camera's pictures.

A movie camera takes a series of still pictures at the rate of 24 per second, or 24 "frames per second." Fortunately for Hollywood, our slow human brains can't assimilate so many separate pictures and we perceive them as all run together, as if the objects in them were progressing smoothly from one position to the next. (Actually, it's our eyes that can't separate the images if they come too close together: Our brains are fast enough. But it still takes my brain more than an hour—if then—to understand what's going on in some movies.)

Let's say that one of the wagon wheel's spokes is painted red. And let's say that when the camera snaps picture number one, the red spoke is pointing straight up, at the twelve o'clock position. Depending on the speed of rotation of the wheel, when picture number two is snapped a twenty-fourth of a second later, the red spoke might happen to be caught in the one o'clock position—even if it had made a couple of complete turns in the interim. That makes it look as if it had moved to the right, or clockwise. Or, it might happen to be caught in the eleven o'clock position, making it look as if it had moved to the left, or counterclockwise. As the camera continues to take its 24 pictures per second, the red spoke—together with the rest of the wheel—will look as if it is moving continuously, either clockwise or counterclockwise.

For extra credit, as we professors like to say, can you figure out how fast the wheel appears to be rotating in this example? (The answer can be found at the end of this section.)

So depending on the number of spokes in the wheel and the actual rotational speed of the wheel compared with the 24 frames per second at which the film was shot, the wheel can appear to be moving forward or backward or—when the spoke speed just happens to be synchronized with the camera's shooting speed—not moving at all. This last is a highly specific coincidence, so it doesn't happen often. But if you look closely, you can see the wheel "stop" briefly as it passes from "forward" motion, when the spoke is slightly ahead of the camera clicks, to "backward" motion, when it is slightly behind.

In reality, of course, the wheels don't have one red spoke; they all look alike. Any spoke is a double for any other. Therefore, any spoke at all might be in the one o'clock or eleven o'clock position when the camera's shutter clicks, and it will still look as if the wheel is turning right or left.

When the wheel is going fast enough, the spokes are moving too fast for the camera's shutter speed to stop their motion. They therefore degenerate into a blur, and the whole effect of backward or forward motion disappears.

You can see exactly the same effects in movies that depict a later mode of transportation: propeller-driven airplanes. When the plane's engine is started, the propeller looks as if it is alternating between the clockwise and counterclockwise directions. As its speed increases, the blades pass through successive "slightly ahead" and "slightly behind" positions with respect to when the camera clicks. As their speed becomes fast enough, the blades become a blur.

Want to see the same effects at home, but you don't have a stagecoach or airplane handy? Try this.

TRY IT　　If you have a portable electric fan, take it into a room that is illuminated with fluorescent light. When you turn the fan on and it speeds up, the blades will appear to be rotating first in one direction and then the other. That's because fluorescent lights flicker on and off 120 times a second (yes, 120; visit the Nitpicker's Corner), and that's five times faster than a projected movie, so we are unaware of the flickering. The "on" flickers are what you see by, so it's just the same as if you were being presented with a series of rapid frames in a movie theater.

NITPICKER'S CORNER

Fluorescent lights (see p. 46) run on alternating current (AC). That means that the electricity flows in one direction for half the time and in the opposite direction the rest of the time. In the U.S., the AC frequency is sixty cycles per second, meaning that one full cycle takes a sixtieth of a second.

Let's say that the current is "positive" for the first half-cycle and "negative" for the other half-cycle. That means that it is "positive" for a hundred-twentieth of a second (half of a sixtieth) and then is "negative" for the next hundred-twentieth of a second, and so on. Thus, there are two current surges (albeit in opposite directions) during each sixtieth of a second, for a total of 120 surges per second. A fluorescent light is "on" only during the current surges, so you might say that it behaves like a movie camera that is snapping 120 pictures per second.

You Didn't Ask, but . . .

Why did everybody look as if they were moving so fast in movies from the early days of the last century (the twentieth, that is)?

Photographic film wasn't as sensitive as it is today, so the exposures had to be longer and therefore further apart in time. The cameras shot only 16 pictures per second, rather than 24. In that longer amount of time between pictures, the people moved farther, so in a second's worth of pictures they seem to have covered more distance. More distance per second equals faster.

Answer to the extra-credit question: There are twelve positions on the clock and the camera is catching the red spoke at the next position every twenty-fourth of a second. The spoke therefore makes one full revolution in twelve twenty-fourths of a second, or half a second. One revolution per half-second is two revolutions per second, or 120 revolutions per minute (rpm).

Damned Spot!

Why does the wet spot on a fabric look darker?

I'll assume that you're in the dining room, concerned about soup on your necktie, although you may have noticed this phenomenon in other rooms under different circumstances.

We see an object because light is coming from that object and entering our eyes. The more light coming from the object, the brighter it appears. And of course, the reverse is also true: An object that is sending less light to our eyes appears darker. So our job is to explain why there is less light coming from the wet spot.

Where does an object get the light that it sends to our eyes? If it is not inherently luminous, like the sun, a lightbulb or Rudolph's nose, then it must be reflecting some of the light that it receives from elsewhere. But nothing reflects *all* of the light that falls upon it; every substance absorbs some light and returns, or reflects, the rest (see p. 38). So the wet spot must be reflecting less light because for some reason it is absorbing more.

Let's take a highly magnified look at the wet fabric as it would be seen by an incoming ray of light.

A fabric is a latticework of interwoven fibers. When it gets wet and soaks up water by capillary action, the spaces between the fibers become filled with water. Many of the incoming rays of light will then be falling upon a water surface instead of striking a fiber.

Now when a ray of light enters a water surface at an angle—and by sheer statistics most of the rays will be hitting the water at an angle, rather than perfectly perpendicular to its surface—a funny thing happens: The ray changes direction. (Techspeak: It is *refracted*. Why does it change direction? Meet me in the Nitpicker's Corner.) Instead of continuing through the water in the direction in which it entered, the light ray veers away from the surface and plunges into the watery depths at an even steeper angle than its entry angle. This steeper angle of penetration means that the light ray penetrates deeper into the depths of the fabric, where it has an increased chance of being absorbed, never to be seen again. Thus, there is more "lost light" inside a wet spot than in a dry one, there is less light reflected and the spot appears darker.

Similar goings-on explain why wet rocks, leaves and grass appear to be more intensely colored when they're wet—why the countryside looks "fresher" after a rain. These objects have colors in the first place because they absorb certain wavelengths of light from the multicolored daylight and reflect the rest back to our eyes (see pp. 34 and 38). When they are coated with a film of water, the incident light rays are refracted deeper into their microscopically rough surfaces.

The refracted light then bounces back and forth off these surfaces, which provides them with many more opportunities for their absorbable wavelengths to be absorbed. The remaining reflected light is thus even more depleted in these absorbed wavelengths than it ordinarily would be, and it therefore looks more intensely colored.

NITPICKER'S CORNER

Why is light "bent" when it enters water?

Whenever a scientist has to explain something about light, he or she has the choice of explaining it on the basis of light *waves* or light *particles* (Techspeak: *photons*), because light behaves as if it were both or either a particle and/or a wave (see p. 170). Explaining refraction on the basis of light's being a wave would require my drawing a diagram and using such terms as "wave front" and "phase velocity," which would make this look too much like (heaven forbid) a science book. So I'll take the easy way out and talk about refraction as if the light ray were a photon bullet.

Better yet, an arrow.

If you stand at the edge of a swimming pool (DO NOT TRY THIS AT HOME!) and shoot an arrow into the water at an angle—not straight down—you won't be surprised to observe that the arrow loses speed as it enters the water and swerves downward, away from the surface. That's because the arrow must travel more slowly in water than in air, and the drag slows down its forward speed. Well, the same thing happens if the arrow is a stream of photons. As they enter the water they slow down and change their direction to a steeper angle than the one at which they entered. The light stream has been refracted. (Note that if you had shot straight down, the arrow would have been slowed, but its direction wouldn't have been changed. It's the same with light; if it enters the water perpendicular to the surface, its direction isn't changed.)

Did I say that light is slowed down when it enters the water? Yes, indeed. But isn't the speed of light always the same? Indeed, no.

When people talk about "the speed of light" as being 186,000 miles per second (3 million kilometers per second), they should always be careful to add "in a vacuum." Because when light enters a transparent medium it slows down, and different transparent media slow it down to different degrees. The speed of light in water, for example, is only three-quarters as fast as it is in air. And that slowing down leads to the bending of the light when it enters water from air.

The bending—*refraction*—of light is even greater when it enters glass from air, because the speed of light in various types of glass is only 50 or 60 percent of its speed in air. Which is just great, because that allows us to use specially shaped pieces of glass—lenses—to *really* bend light a lot and make all sorts of clever gadgets such as telescopes, microscopes and eyeglasses.

Spurn That Burn

My dermatologist told me that a sunscreen lotion labeled SPF 30 does not block out twice as much harmful radiation as one labeled SPF 15. What gives?

Your doctor is correct. The SPF numbers aren't sun-*filtering* factors—they're sun-*protecting* factors. SPF stands for "sun protection factor." The numbers are not telling you how much radiation they block out, but how much *time* you can spend in the sun before your skin turns red, a condition doctors call erythema. And that's quite another matter.

With an SPF 15 on you, you can stay out in the sun fifteen times longer than with bare skin. With an SPF 30, you can stay out thirty times longer than with bare skin. That's twice as long as with an SPF 15. And yet an SPF 30 blocks out only about 3 percent more of the harmful radiations than an SPF 15 does!

I'm well aware that the foregoing is probably the most confusing paragraph you have ever read outside of an IRS publication. But I'll show you that it's all quite logical.

First, though, what are those menacing radiations that rain down upon us from our life-giving star? The sun's atoms, being as hot as they are (about 9800 degrees Fahrenheit or 5400 degrees Celsius at the sun's surface), are continually giving off radiations of almost every energy . . . uh, under the sun, ranging from radio waves to X rays. The dangerous X rays are pretty much filtered out by Earth's atmosphere, while the sun's radio waves are substantially less harmful than those emanating from a hard-rock radio station. That leaves only visible light and two types of invisible radiations: infrared (see p. 112), which warms us but doesn't burn us, and ultraviolet. This last one is the villain.

Ultraviolet (UV) radiation is usually subdivided into three regions of energy, which scientists have imaginatively labeled A, B and C. We can eliminate ultraviolet C (abbreviated UVC) from our fears, because it is absorbed by the atmosphere's ozone layer, which, though threatened by human activities, is still pretty much up there. So the only things we have to worry about at the beach besides our paunches and cellulite are UVA and UVB, which can cause not only sunburn, but permanent skin damage and cancer.

Sunscreens are a mixture of active chemicals in a cosmetically appealing base. The molecules of any chemical selectively absorb radiations of specific energies (see p. 37). The sunscreen chemicals have prodigious appetites for absorbing ultraviolet radiation, even when in extremely thin layers on the skin. On the labels of sunscreen containers, you'll see UVA absorbers such as avobenzone or Parsol; UVB absorbers such as octyl methoxycinnamate and other cinnamates, homosalate, octyl salicylate and padimate O; and double-threat UVA-UVB absorbers such as oxybenzone and other benzophenones. A chemical called PABA used to be popular, but it irritated some people's skin and is no longer used.

Okay, chemistry class dismissed. But I thought you'd like to be able to interpret the ingredient lists on the product labels.

Not to worry about the names, however. Most products are carefully balanced witches' brews of chemicals designed to absorb the entire range of harmful UV energies. But remember that they are tested primarily for burn prevention, whereas research continues to find certain UV energies to be worse than others at causing premature skin aging or cancer. It's best to choose a "broad spectrum" sunscreen to cover both your back and your bets.

Now back to those tricky SPF numbers. It's all in the arithmetic. Watch me. Nothing up my sleeve.

Suppose that Brand X sunscreen cuts out half—50 percent—of the burn-producing UV rays. Obviously, you could stay out twice as long as usual without burning. If you'd ordinarily burn in one hour with no protection, you could stay out for two hours. In other words, the SPF is 2.

Now suppose that Brand Y cuts out 75 percent of the UV rays, which means that you're being exposed to only 25 percent of the burning rays instead of 100 percent. You'd be able to stay out four times as long as with no protection, wouldn't you? ($100 \div 25 = 4$.) The SPF then is 4. *Brand Y cuts out only 25 percent more of the UV rays than Brand X does, yet its SPF is twice as high: 4 instead of 2!*

I won't go through the algebraic derivation (do I hear a release of bated breath?), but if you want to figure out the percent of absorbed burning rays from an SPF number, here's how: subtract 1 from the SPF, multiply by 100, and divide the result by the SPF. For example, for an SPF of 20: $20 - 1 = 19$; times $100 = 1,900$; divided by $20 = 95$ percent absorption.

In that way, you can figure out that an SPF of 15 absorbs

93.3 percent of the UV rays, while a twice-as-big SPF of 30 absorbs 96.7 percent, only 3.4 percent more.

You see that by paying more money for a higher-SPF product, you're blocking only a small amount of additional radiation. It's a classic case of diminishing returns. Even if you're a creamy-skinned redhead whose skin tends to match your hair after an hour in the sun, you don't really need an SPF of more than, say, 30. What makes you think you're going to be outdoors for more than thirty hours, anyway? The sun does have a habit of setting, you know.

BAR BET A sunscreen rated at SPF 30 allows you to stay out in the sun twice as long as an SPF 15, yet it cuts out only about 3 percent more of the sunburning rays.

Wrong, Wrong, Wrong!

Those "light windmills" that we see spinning around in the windows of novelty stores: What makes them work?

They're called radiometers and are generally supposed to illustrate that light has pressure. But they don't. If a machine could be a con artist, this gadget would take the cake.

You've seen them. They look like a lightbulb on a stand. Inside the bulb, which has had most of the air pumped out, are four thin, metal vanes, mounted like a pinwheel on a low-friction pivot. One side of each vane is shiny (or sometimes white), while the other side is black. The shiny side of one vane faces the black side of the next, and so on. When exposed to sunlight, the vanes spin merrily around, away from their black sides and in the direction of their shiny sides.

People have been trying to find out what makes the radiometer turn ever since 1873, when it was invented by Sir William Crookes (1832–1919). He thought it was pressure from the light, which was somehow pushing harder on the black surfaces than on the shiny surfaces. Sir William, who was a smart man but was wrong about the light-pressure effect, launched a scientific quest that hasn't stopped yet. Even today's encyclopedias give a popular, but demonstrably wrong, explanation of how Crookes's radiometer works.

Warning: You are about to encounter one of only two places in this whole book (I hope) in which the answer to a question will be somewhat less than satisfying. The best current explanation of the radiometer, which I promise I will give you at the end, is a bit hard to swallow and is still being doubted by some scientists, including me. The other less-than-satisfying explanation is why the shower curtain is sucked inward during your shower. (For that one, see p. 188.)

First, let's debunk some of the obviously wrong radiometer explanations that are circulating as recklessly as a radiometer in hell.

Radiation pressure

Light, as everyone knows, is electromagnetic radiation. And electromagnetic radiation, as you either know or can quickly find out (see p. 170), is a stream of tiny packages of energy called photons. Photons act like little bullets, insofar as when they hit something, they can have a physical impact. For example, light photons can actually knock electrons out of many solid substances. That's called the photoelectric effect, and the photon explanation that I just gave you won Albert Einstein a Nobel Prize. (He explained it in a little more detail than I just did.)

So, one might think, it's the stream of photon bullets hitting the radiometer's vanes that spins them around, just like when you—DO NOT TRY THIS AT HOME!—shoot a machine gun at a weather vane. While radiation pressure does indeed exist, we now know that it is much too weak to be pushing those vanes around. Moreover, radiation pressure should make the radiometer turn the other way!

Here's why. Light is absorbed by black surfaces (see p. 38) and reflected by shiny surfaces. The black surfaces of the vanes simply swallow the photons, whereas the shiny surfaces spit them right back out again (see p. 34), getting a backward recoil kick just as a gun gets when spitting out a bullet. That would make the vanes spin *away* from their shiny sides and *toward* their black sides—just the opposite of what we see happening.

Gas pressure

This appears to be the best-loved of all the wrong explanations. It is dispensed by the *Encyclopædia Britannica* and other encyclopedias, as well as by many science teachers.

The story goes that the black surfaces of the vanes absorb more light energy than the shiny sides do and are thereby slightly warmer. (Correct so far.) The air adjacent to the black sides—there's still a small amount of air in the bulb— is warmed by this energy (still correct), which makes the air pressure higher on the black sides (wrong!). This supposedly increased pressure pushes on the black sides, making the vanes move toward their shiny sides.

But let's ask the following question: When the air is heated, which indeed makes its molecules move faster, why should those faster-moving molecules dash themselves against the vanes any more often than they dart off in any other direction? There can be no net directional force from the molecules' motion. Putting it another way, the air's pressure *can't* increase, because it is not confined. It is free to expand and relieve any incipient pressure anywhere it likes within the bulb, so there is no more reason for it to expand against the vanes than in any other direction. Thus, there is no net vane-pushing force caused by the warmer air.

Outgassing

Some conspiracy theorists would have us believe that the black coating on the vanes contains *adsorbed* (surface-bound) gases, and that when the black sides are heated by absorbing light, those gas molecules are expelled, sort of like popcorn from a frying pan. The leaping gas molecules would exert a force on the black surface, just as a basketball player exerts a force on the court floor when he jumps, and this force pushes the vanes around. But if this were true, the radiometer would eventually wear out as all the adsorbed gases were released.

Photoelectric effect

What if the photons of light are ejecting electrons from the black sides of the vanes and, in departing, the electrons give a backward kick to the vanes? No cigar on that one, either, because you can make radiometer vanes out of materials that don't exhibit the photoelectric effect; their electrons are held too tightly for visible light to be able to knock them out. Also, the photoelectric effect would still occur even if the bulb were completely evacuated, but the radiometer won't work without some air in the bulb.

Convection currents

The heated black surface sets up air currents by convection (see p. 112), and the moving air blows the vanes around. The only trouble with this one is that nobody can invent any air currents that blow mainly in one direction: against the black sides of the vanes.

The best scoop

In 1881, a British mechanical engineer named Osborne Reynolds (1842–1912) published a paper that explained the radiometer in a way that many scientists now grudgingly accept. The reason it isn't more widely known is probably that it isn't easy to describe or to understand. But here goes.

It has something to do with the temperature difference between the warmer air adjacent to the black sides of the vanes (due to their energy-absorbing nature) and the cooler air adjacent to the shiny sides. Apparently, when this air flows out to the edges of the vanes, the warmer, faster molecules strike the edges at a more oblique angle than the cooler molecules do, and that pushes the vanes in a direction away from the black sides. Exactly *why* this should be true is buried in complex mathematics, which I shall not attempt to decipher for you (or me). I confess that it's hard for me to believe that it's the edges of those skinny vanes, rather than their broad surfaces, that push them around. But that's what Mr. Reynolds says, and none of the other explanations stands up under close examination.

I warned you, didn't I?

You Didn't Ask, but . . .

If scientists today can unravel the mysteries of life itself, why can't they explain the simple little radiometer after more than a hundred years of trying?

The main answer is that they haven't really been trying. There has been no vast federal program to inject billions of dollars into radiometer research as there have been for the Manhattan (atomic bomb) Project, the space program, genetic research and other health-related enterprises. Not that money alone can solve a scientific problem, but scientists are like everybody else: They tend to do what they get rewarded for, and nobody is going to get a research grant, a promotion or a Nobel Prize for figuring out how a toy works.

Window, Window, in the Wall, How Come You Block No Light at All?

Why are air, water and glass transparent, when practically no other materials are?

Well, what does "transparent" mean? It means that any light being reflected in our direction from an object outside a glass window, for example, can pass right through the glass unobstructed and come out the other side, where our eyes can deal with it. We therefore *see* the object *through* the window. That's why people who have little regard for the English language use the term "see-thru" (invariably spelled that way) instead of the perfectly good word *transparent.*

In general, when a traveling ray of light encounters a new substance, it may be *reflected* backward from the surface or it may penetrate the surface and be *absorbed.* If it manages to escape both of these fates, it can continue traveling through the medium; it will be *transmitted.* So our job is to explain why air, water and glass don't reflect and/or absorb very much of the light they receive. Almost all other substances—except some waterlike liquids and glasslike plastics—absorb some of the light and reflect most of it (see pp. 34 and 38), leaving practically none to be transmitted.

Let's get air out of the way first. Under ordinary conditions, the spaces between air molecules are around ten times bigger than the molecules themselves. So air is almost completely empty space, containing virtually nothing that could interfere with the passage of light except for a very occasional molecule. Ditto for all gases.

Water and glass are quite a different ball game, however, because their molecules are very close together—close enough to do a fair amount of reflecting. Remember the glare from that pond's surface or from that car's windshield on a sunny day? So even from the most transparent liquid or solid substances, some light is reflected. It depends on the angle at which the light hits the surface.

Of the light rays that do succeed in penetrating air, water or glass, very, very few of them are absorbed; almost all the light gets through. Molecules absorb light because their electrons have certain preferred energies (see p. 34), and by taking on the extra energy of a light particle (a photon; see p. 170), they can reach another, higher one of their preferred energies. It happens that none of the molecules in air, water or glass can absorb and "use" any of the energies in visible light; the energies that they *can* absorb are certain radiations that humans can't see, such as ultraviolet and infrared radiations. Ditto for alcohol, kerosene and other familiar transparent liquids. So if very little light is absorbed and the angle isn't right for reflecting, almost all of the light will go straight through by default.

NITPICKER'S CORNER

There are, of course, *colored* glasses, liquids and even gases. What's going on there is that they selectively absorb *some* of the wavelengths or energies in white (or colorless) light (see p. 38) and transmit only those that they can't "use." The transmitted light therefore has a different composition of wavelengths from white light and hence a perceived color.

You Didn't Ask, but . . .

Why is a mirror such a good reflector?

Mirrors are the best reflectors of light that human ingenuity has been able to devise. Notice, however, that the light is reflected only from the *backing* of the mirror, after passing through the front layer of transparent glass.

What is there about the backing that makes it such a good reflector? It's a thin, smooth layer of silver metal. All metals are shiny, or reflective, because their atoms are held together by a sea of loose, swarming electrons that have no affiliation with any particular atoms. (That's why metals conduct electricity so well—because electricity is just a movement of electrons.) The swarm of footloose electrons in the silver, belonging as they do to no particular atoms, have no particular preference for absorbing any specific wavelengths of light (see p. 38), so they reject and reflect back all wavelengths.

Of course, a sheet of shiny silver metal would make a fine mirror without the glass, but it would quickly tarnish.

A Light Bite

Why do WintOGreen Life Savers make flashes of light?

Your question may sound silly to those who haven't heard about it before, but chomping on those little candies really does make flashes of light. It may not help you at all to know that the phenomenon is called *triboluminescence,* but there, I've said it and done my duty as a scientist.

Life Savers, it will not surprise you to know, are little more than donut-shaped crystals of sugar. Certain crystals, including cane sugar, have long been known to exhibit this property of tri . . . whatever. In fact, way back in 1605, the English philosopher Sir Francis Bacon (1561–1626) reported that when he chopped up blocks of sugar in the dark (sugar was sold in big blocks and candlelight was dim), he observed flashes of "a very vivid but exceedingly short-lived splendour." Mineralogists have long known that certain mineral crystals also give off light when subjected to sudden shock.

Here's what's going on.

A crystal is an orderly, geometric arrangement of atoms, all bound together into a sort of three-dimensional latticework structure. Examples that you may be familiar with are sugar (sucrose), salt (sodium chloride), quartz (silica) and diamond (an overpriced form of carbon). It has been found that crystals whose molecular arrangements are not symmetrical—that is, whose molecules are not situated identically in two opposite directions—are the best flashers.

When such a crystal is cracked open, the atoms are torn

apart from one another and some of their electrons are torn off in the process. Crystal fragment A may wind up with more electrons than it deserves, while crystal fragment B may not have enough. As they begin to separate, the extra electrons on fragment A are attracted strongly back to where they belong, and they zap across the widening air gap between A and B, exactly like a bolt of lightning zapping through the air between a cloud and the ground.

These miniature lightning bolts make tiny blue flashes because the air's molecules are energized by the swift rush of electrons through them, following which they throw off their extra energy in the form of light (see p. 34). Hard, fracturing whacks on the crystal can therefore produce weak flashes of light.

That's all that happens in most triboluminescent crystals. But in the case of WintOGreen Life Savers, that's *not* all that happens. There's an almost instantaneous second step that makes the light much brighter.

Much of the "lightning" that the electron-zapped air gives off is invisible to humans; it is ultraviolet radiation, which is of higher energy than visible light. But WintOGreen Life Savers contain a chemical called methyl salicylate, also known as oil of wintergreen; it's the flavor in the leaves of the wintergreen plant, a small, creeping evergreen sometimes known as teaberry. This chemical has the property of being *fluorescent* (see p. 34). That is, its molecules absorb the ultraviolet radiation (see p. 34) and re-emit it as visible light. It's that visible light that is mainly what you see when someone chomps a WintOGreen Life Saver in a dark closet.

Can't wait to try it?

TRY IT Take a roll of WintOGreen Life Savers into a dark closet with a hand mirror or a close friend. (If you're already in the closet, so much the better.) Make sure the closet is completely dark; wait until

nighttime and plug the crack under the door with a towel if necessary. Think pure thoughts for about ten minutes, while your eyes become thoroughly dark-adapted. Now pop a Life Saver into your mouth and, in spite of what your mother taught you, quickly crunch it noisily between your teeth with your mouth open. Your mirror or your friend will see surprisingly bright flashes of light inside your mouth. Baby dragons are trained on WintOGreen Life Savers.

You may also want to play around with sugar cubes. In the closet, clash them glancingly against each other, as if striking a match. You'll see the miniature lightning flashes, but they won't be brightened by the fluorescence of wintergreen's methyl salicylate.

3

Hot Stuff

Everything is hot. That is, it contains some heat. And as a consequence, it has a temperature. Even an ice cube contains heat. "Hot" is strictly a relative term.

Heat is the ultimate form of energy, the form into which all other forms ultimately degenerate. There is energy of motion (Techspeak: *kinetic energy*), there is gravitational energy, chemical energy and electrical energy. They can all be converted into one another with the right equipment. We can convert gravitational energy into kinetic energy by pushing a boulder off a cliff. We can convert a waterfall's kinetic energy into electrical energy by connecting a waterwheel to a generator. We can convert chemical energy into electrical energy with a battery, and so on.

But no conversion can be 100 percent complete. Some of the energy must inevitably be "wasted"—turned into heat. When the boulder hits the ground it heats it up a bit and we lose that amount of heat energy. When the waterwheel turns, its bearings get warm from friction and we lose that amount of heat energy. When a battery delivers current it gets hot from the chemical reactions inside and we lose that amount of heat energy. In short, we can convert and reconvert energy as much as we like, but each time, we will lose a little in the form of heat.

Can we collect that "wasted" heat and convert it back into another form? After all, we seem to be recycling everything these days; can't we recycle heat energy? Sure, but not completely. That's because heat is a chaotic motion of atoms and molecules (see p. 79), and to restore them to order takes work: energy. We must spend energy to recover that heat energy, so the bottom line on the energy balance sheet will always show a net deficit.

The preceding ideas are embodied in what is known as the Second Law of Thermodynamics, which is one of the most profound sets of realizations ever to dawn upon the mind of man.

But although we can't use it with 100 percent efficiency, heat is far from a minor player in the energy game. The world thrives on heat. It is the common currency, the euro of energy, if you will, that we humans manipulate to suit our energetic objectives. We add it to our ovens and we remove it from our refrigerators—after first converting it into electricity, of course, which is so much easier to handle than fire.

Like unfettered physical objects, heat can travel from one place to another as long as it is going "downhill": from someplace at a higher temperature to someplace at a lower one. In that sense, flowing heat is very much like flowing water.

But does the heat flow because the higher-temperature object contains more heat than the lower-temperature object? Not necessarily. People often confuse heat with temperature—people who haven't read this chapter, that is.

Using water flow as an analogy to heat flow, try this riddle on for size. Then return to it after you've read the section that begins on page 79.

> *If a waterfall flows spontaneously down from lake A into lake B, does that mean that there is more water in lake A than in lake B? (Note: Heat is analogous to the amount of water, while temperature is analogous to the altitude.)*

This chapter, then, is about heat and the electricity that we make out of it. It's about global cooling (yes, cooling), cold feet, cold steel, hot fire, sparrows, refrigerators, thermometers and bathtubs.

Who is this guy Lewis Carroll, with his shoes, ships and sealing wax?

Double Trouble

I live in Miami and my twin sister lives in Tucson. One day on the telephone I mentioned that it was 80 degrees Fahrenheit (26 degrees Celsius) in Miami, and she jokingly said that it was "twice as hot" in Tucson. If that were really the case, what temperature would it be in Tucson?

It certainly wouldn't be 160 degrees Fahrenheit (71 degrees Celsius). But that's not because 160 degrees is too hot; it's not hot enough. The temperature that is "twice as hot" as 80 degrees Fahrenheit, believe it or not, is 621 degrees Fahrenheit!

Here's what's going on.

First of all, we must realize that heat and temperature are two different things. Please repeat after me: *Heat is energy, while temperature is just our human way of telling one another how densely concentrated that heat is in an object.*

Let's take heat first.

The amount of heat energy an object contains can be counted up in calories, just as if it were a donut. (A calorie is just an amount of energy, right?) But you'll grant that a big donut contains more calories than a small donut, won't you? Well, it's the same with the energy content of any substance. A quart (a liter) of boiling water contains twice as much heat energy as a pint (half a liter) of boiling water, even though they're both at the same 212-degree temperature.

Another example: There's a lot more heat in a bathtub full of warm water than in a single glassful that you might scoop out of that same bathtub, simply because there are more hot molecules in the tub. In short, the more of a substance you have, the more heat energy it contains.

(Right here, you may wish to take time out to take a crack at the riddle on page 78. The answer can be found at the end of this section.)

So your sister's problem, whether she realized it or not, was to figure out how much heat there actually was in the outside air—let's say a cubic yard (or a cubic meter) of it. Then if there was twice as much heat per cubic yard (or cubic meter) in the outside air as in your Miami air, she could really say it was "twice as hot."

How can we determine the amount of heat in an object? Taking its temperature won't do the job, because that doesn't account for how big the object is. As we discovered in the bathtub, a big object containing lots of heat can be at the same temperature as a smaller object containing much less heat. Moreover, temperatures, whether expressed as Fahrenheit or Celsius, are nothing but arbitrary numbers invented by those two eponymous gentlemen (see p. 87). They're merely convenient labels for people to talk about—numbers that everyone has agreed to, as if proclaimed on Mount Sinai: *"Whensoever thine ice melteth, it shall be called 32 degrees Fahrenheit or zero degrees Celsius. And whensoever thy water boileth, it shall be called 212 degrees Fahrenheit or 100 degrees Celsius."* These proclamations were made not by the Lord, but by Messrs. Fahrenheit and Celsius.

But the amount of heat that an object contains cannot be subject to humans' monkeying around with numbers. We need an absolute way of expressing the heat content of things.

The crux of the problem is that on either of our temperature scales, zero temperature does not mean zero content of heat. Zero degrees Celsius, for example, is merely the temperature of melting ice. Does that mean that there can be nothing colder than melting ice? Of course not.

Or look at it this way: How can you use a scale to measure something if its zero doesn't really mean zero? Picture a yardstick (or meter stick) with "zero inches" (or "zero centimeters") somewhere in the middle, instead of at the left end. Just think of the crazy measurements you'd get.

So if we're ever going to be able to measure the amount of heat in an object, or in the air for that matter, we'll have to have a scale of numbers on which zero actually means no heat at all. And that's where the Lord really does come in. No, not that Lord. Lord Kelvin, a British nobleman and scientist (1824–1907), whose street moniker was William Thomson.

Kelvin set up a scale of temperatures that begins at "no heat at all"—a temperature of absolutely zero, where things are as cold as they can possibly get: "absolute zero." Then, he borrowed the size of Mr. Celsius's degree and started counting upward from there. When you do that, the temperature of freezing water, zero degrees Celsius, turns out to be 273 degrees above absolute zero, and the temperature of boiling water—100 degrees Celsius—is 373 degrees above absolute zero. Human body temperature (37 degrees Celsius) turns out to be 310 degrees on the absolute scale. (Tell *that* to your doctor when he asks what your temperature is.) You can see that the absolute temperature, measured in Kelvins in honor of Lord Kelvin, is the Celsius temperature plus 273.

Now we're ready to answer your sister's riddle. If Tucson air contains twice as much heat per cubic yard (or per cubic meter) as Miami air, then what we must double is the *absolute* temperature of the Miami air. First converting your 80-degree Fahrenheit temperature into Celsius (for how to do that, see p. 90), we get 27 degrees Celsius. Adding 273 gives us 300 Kelvins, which is now a real measure of the heat content of the air. Doubling it to get twice the heat, we get 600 Kelvins, which converts to 327 Celsius or 621 degrees Fahrenheit as your sister's offhand estimate of the Tucson temperature! Yeah, we know, Sis: You don't feel it because the humidity is so low, right?

Similarly, inside your house, if your thermostat is set on 70 degrees Fahrenheit and you want twice the heat, you'd have to turn it up to 599 degrees Fahrenheit. You can take my word for that or do the calculation yourself. In a Celsius country, you'd have to turn the thermostat up to 313 degrees in order to make a 20-degree house twice as hot.

BAR BET If your room's temperature is 70 degrees Fahrenheit and you want it to be twice as hot,

> you'd have to turn your thermostat up to about
> 600 degrees Fahrenheit.

To win this bet, there's no need to scribble calculations on a napkin. Just point out politely to your buddies as you pick up the money that the Fahrenheit scale doesn't read zero when there's a complete lack of heat; there's an awful lot of unaccounted-for heat *below* zero. Thus, the number "70" doesn't account for all the heat, and doubling it won't get you anywhere near twice the amount of heat.

You Didn't Ask, but . . .

Why is there a limit to how cold anything can get?

Heat is energy.

What kind of energy?

It's not electrical energy or nuclear energy or the kind of energy that your car has as you barrel down the highway. It's the energy that an object contains *within* itself, because the particles that it's made of, its atoms and molecules, are actually vibrating and bouncing around within their limited spaces like a bunch of maniacs in padded cells. The more vigorously those particles are moving, the hotter we say the stuff is: the higher its temperature. Even at the same temperature, though, a bigger chunk of the stuff will contain more heat energy because it contains more moving particles.

When we cool something down by taking heat energy out of it, the energy is lost by those moving particles, which will then be moving more slowly. Ultimately, if we cool it down far enough, we should reach a point where the particles stop moving altogether. We will have reached the lowest possible temperature: absolute zero.

And by the way, when you want to tell your doctor that you have no fever, please don't say that you have "no temperature." That would mean that your body is at absolute zero, in which case a doctor would be of no help whatsoever.

NITPICKER'S CORNER

Atomic and molecular motion don't stop dead-still at absolute zero. Theory says that there would be a tiny bit of residual energy left. But absolute zero isn't based on molecular motion anyway. It's the temperature at which a gas would shrink so much from the cold that it would disappear entirely. Nobody has yet succeeded in cooling a substance down to precisely absolute zero—in fact, theory says that it can never actually be reached—although experiments have gotten to within several billionths of a degree of it. For one thing, you'd have to keep a substance inside an absolute insulator, through which not a single atom's worth of heat can penetrate. And that's not exactly a job for a Kmart thermos bottle.

Answer to the riddle on page 78: Of course the waterfall doesn't care how big the lakes are. But just as water will flow from a higher altitude to a lower one no matter how much water is involved, heat will flow from a body at a higher temperature to a body at a lower one, no matter how big or small the bodies may be or how much actual heat they therefore contain. It's the temperature *difference* that counts—the difference in energy between the fast, hot molecules and the slower, cooler ones that they collide with and transmit their energy to.

Cold Feet

Why does the tile floor in my bathroom feel so cold on my bare feet?

Assuming that you haven't forgotten to pay your gas bill, it's because porcelain tile conducts heat better than that cozy bath mat does, even though they're at the same temperature.

It's a common experience that certain things feel colder than others. People talk about "cold steel" as if the blade of a sword were somehow colder than its surroundings. Bakers like to roll out their pastry dough on a marble slab because "it's colder." Just touch a steel knife blade or a marble slab and you'll have to agree: They *do* feel colder.

But they're not. *The steel, the marble and the floor tiles aren't one bit colder than anything else in the room.* They just feel as if they are.

If they have been in the same room for any reasonable amount of time, all objects will be at the same temperature as everything else in the room, because temperatures automatically even themselves out. Hot coffee cools off and cold beer warms up. Let a cup of hot coffee and a glass of cold beer stand side by side on the table long enough and they'll eventually come to the same temperature, the prevailing temperature of the room. (Nevertheless, you'll still think of the coffee as "cold" and the beer as "warm," won't you?)

The reason is that heat spontaneously flows from warmer to cooler. That's because the molecules in a warm object are moving faster than the molecules in a cool object; that's what temperature *is:* a measure of the molecules' average speed. So when a warm object comes into contact with a cool one, its faster molecules will collide with the slower molecules and speed them up—that is, make them warmer.

If an object should happen to be initially colder than its surroundings, then heat will automatically flow into it until its temperature is the same as its surroundings. Or if an object happens to be initially warmer than its surroundings, heat will flow out of it into the surroundings. We have found it useful to think of heat flow as if it were flowing water. Water always flows to a lower level, while heat always flows to a lower temperature (see p. 79). We might even say that temperature seeks its own level.

It's not just steel, marble and tile that will feel cool to your touch. The temperature of your skin is a bit below 98.6 degrees Fahrenheit (37 degrees Celsius), while everything else in your room (except a hot radiator, perhaps) is at the room's prevailing temperature—around 70 degrees Fahrenheit (21 degrees Celsius). So when you touch an object in the room, it will feel cool to your skin because it really *is* cooler than your skin. Heat will therefore flow from your skin into the object, and your heat-deprived skin gives you the sensation of coolness.

TRY IT Pick up various objects in the room you're in, other than obviously cool or warm objects such as a bottle of pop, a cup of coffee or the dog. Press them one by one against your forehead. They will all feel slightly cool to you.

But as you have discovered to your discomfort, some things do feel colder than others; the tile floor feels colder

than the bath mat, even though we've seen that they must be at the same temperature. How come?

The answer is that, while all objects in the room are cooler than your skin and will therefore steal some heat from it, some materials are better heat thieves than others. Some materials are better *heat conductors*—they are better at carrying the stolen heat away. And the faster a material conducts the heat away, the cooler your skin is going to feel. It happens that porcelain tile is a much better heat conductor than the cotton or synthetic fiber bath mat is, so heat flows faster out of your tootsies when they're on the bare floor and they feel colder.

Different substances, being made of molecules with different properties—and that, of course, is *why* they are different substances—will transmit this heat with different degrees of speed and efficiency.

Substances made of big, unwieldy or rigidly fixed molecules won't be able to jostle their neighbors as easily, so they won't be able to transmit heat as quickly. That's the case with substances like cotton, wood and rubber, for example. Your wooden floor doesn't feel as cold as the tile floor, does it? That's because its big molecules can't steal heat away from your skin as fast.

Among all types of materials, gases are the worst conductors of heat. Their molecules are so far apart that they can barely find other molecules to bump up against. Almost everything conducts heat better than air, and that's why almost everything you touch feels cool to some extent; you're comparing it with the air that you're normally surrounded by and accustomed to. Air insulates you.

Metals, on the other hand, are the best conductors of heat among all materials, because of their unique structure. They contain loose electrons that can drift easily from one atom to another. That's why metals conduct electricity so well, but it's also why they conduct heat so well. Those tiny electrons can carry heat energy from one place to another much more efficiently than big, jostling atoms or molecules can, because they're so much more mobile. As the best thieves of heat from your skin, metals will feel coldest of all.

Engineers and physicists have measured how well many substances conduct heat (Techspeak: their *thermal conductivities*). Here, in round numbers, is how the thermal conductivities of some familiar materials stack up compared with air, to which I've assigned the number 1. The farther down the list you go, the colder the substance will feel when it's really at room temperature.

HOW VARIOUS MATERIALS CONDUCT HEAT, RELATIVE TO AIR

Material	Value	Material	Value
Air	1	Granite	130
Rubber	6	Stainless steel	600
Wood	6	Iron	3,300
Water	24	Aluminum	9,500
Glass	30–40	Copper	16,000
Porcelain tile	40	Silver	17,000
Marble	70–120		

The moral of the story: Never build a house with silver bathroom floors. Or toilet seats.

Gabriel, You Blew it!

Why are the temperatures of freezing and boiling such odd numbers as 32 and 212 degrees Fahrenheit?

They are indeed strange numbers for such common, everyday goings-on as the freezing and boiling of water. We're stuck with them because a German glassblower and amateur physicist named Gabriel Fahrenheit (1686–1736) made a couple of bad decisions.

Gadgets for measuring temperature had existed since about 1592, even though nobody knew what temperature was, and nobody had tried to attach numbers to it.

Then in 1714 Fahrenheit constructed a glass tube containing a very thin thread of mercury—a nice, shiny, easily visible liquid—that went up and down by expansion and contraction as it got hotter and colder. But Fahrenheit's thermometer, like all that preceded it, was like a clock without a face. He had to put numbers on the thing, or else how could anyone complain about the weather?

So Fahrenheit had to devise a set of numbers to inscribe on his glass tubes, such that the mercury would rise to the same number on all thermometers when they were at the same temperature. And that's when Gabriel blew it. Historians still speculate about what must have been going through his mind, but the following might be a good guess.

First, he decided that because a full circle has 360 steps or degrees, it would be nice if there were also 360 steps—and why not call them degrees—between the temperatures of freezing water and boiling water. But 360 steps would make each degree too small, so he chose 180 instead.

That fixed the size of the degree: exactly $1/180$th of the distance on the tube between the freezing and boiling marks. But what, he wondered, should the actual numbers be? Zero and 180? 180 and 360? Or, heaven forbid, 32 and 212? (212 − 32 = 180, right?)

Well, he stuck his thermometer into the coldest concoction he could make—a mixture of ice and a chemical called ammonium chloride—and called that temperature "zero." (What arrogance, Gabriel! Would nobody in human history ever be able to make a colder mixture? Why, two centuries later, we can make temperatures almost 460 degrees below your zero.)

When he took his own temperature, the thermometer went up to around 100 degrees. (Okay, 98.6, but see the following for how that number came about.) That was a touch that Fahrenheit liked: Humans, he felt, should score 100 on his temperature scale.

Next, he stuck his thermometer into an ice-water mixture, and found that the mercury went up 32 degrees higher than in his zero-temperature mixture. And that's how the freezing point of water came to be 32 degrees. Finally, if boiling water was to be 180 degrees higher than that, it would wind up at 32 + 180, or 212. End of Gabriel Fahrenheit's story.

Six years after Fahrenheit's body temperature became equal to that of his surroundings, a Swedish astronomer named Anders Celsius (1701–1744) proposed the centigrade scale of temperature, which we now call the Celsius scale. "Centigrade" means 100 degrees; he set the size of a degree so that there are 100 of them, not 180, between the freezing and boiling points of water. Furthermore, he defined his "zero temperature" at the freezing point of water, a reference point that anyone could easily reproduce. And thus, the boiling point of water fell at 100 degrees. (Curiously, for reasons known only to eighteenth-century Swedish astronomers, Celsius originally took the freezing point as 100 and the boiling point as zero, but people turned it around after he died.)

And what about that number 98.6 as the "normal" human body temperature? It's just a fluke. People's temperatures vary quite a bit depending on the time of day, the time of month (for women) and just plain differences in metabolism. But it wavers around an average of 37 degrees Celsius for most people, so that's what doctors have adopted as "normal." And guess what 37 degrees Celsius converts to in Fahrenheit? Right—98.6, a number that looks for all the world as if it were more precise than it really is. That extra six-tenths of a degree is nothing but an accident of the conversion arithmetic and has no significance at all.

Speaking of conversions, I can't resist the opportunity—I do it every chance I get—to publicize an *easy* way to convert temperatures. I don't know why they continue to teach those complicated formulas in school, with all their 32s, parentheses and improper fractions, when there is a much simpler way that's absolutely accurate.

Here's how:

> *To convert Celsius to Fahrenheit, add 40, multiply by 1.8, then subtract 40.*
> *To convert Fahrenheit to Celsius, add 40, divide by 1.8, then subtract 40.*

That's all there is to it. It works because (a) 40 below zero is the same temperature on both scales and (b) a Celsius degree is 1.8 times larger than a Fahrenheit degree. (180 ÷ 100 = 1.8.)

A final point: *Thermometers measure only their own temperatures.*

Think about it. A cold thermometer registers a low temperature; a hot thermometer registers a high temperature. A thermometer doesn't register the temperature of an object that you stick it into until it itself warms up to, or cools down to, that object's temperature. That's why you have to wait for the fever thermometer to warm up to your body's temperature before you read it.

BAR BET A fever thermometer doesn't measure your body's temperature; it measures its own temperature.

Hot, Hotter, Hottest?

If absolute zero is the lowest possible temperature, is there a hottest possible temperature?

Yes. But let's start off at merely warm and gradually turn up the heat.

Heat is the energy that a substance contains within itself, due to the fact that its atoms and molecules are moving (see p. 79). But temperature is a man-made concept, invented so that we can converse among ourselves about how much of that energy a substance has and actually assign numbers to it. When we say we are "raising the temperature" of an object, we are adding heat energy to its atoms and molecules and making them move faster. The ultimate limit to cooling and slowing them down has to be when they're not moving at all; that's absolute zero (see p. 79). Our current question, then, comes down to whether there is any limit to how fast those atoms and molecules can move.

But long before we reach any such speed limit, several things will happen. First, if the substance is a solid it will melt into a liquid. Then at a higher temperature the liquid will boil and become a vapor or gas—a condition in which the atoms or molecules are flitting around freely in all directions. As the temperature gets higher and higher, they flit faster and faster. For example, the nitrogen molecules in the air in your 350-degree-Fahrenheit (177-degree-Celsius) oven are flitting about at an average speed of 1,400 miles per hour (2,300 kilometers per hour).

If the substance is made of molecules (clusters of atoms glued together), the molecules will eventually be knocked to pieces—broken apart into smaller fragments or even into their individual atoms by the shattering forces of their violent collisions. In other words, every molecular compound will decompose at a high enough temperature.

Will the individual atoms themselves ever be broken apart? Yes, indeed. At a high enough temperature the atoms' electrons will be torn off, resulting in a seething, fluid inferno of free electrons and charged atomic fragments, called a plasma. This is the stuff of the interiors of stars, at temperatures in the tens of millions of degrees.

Still higher temperatures? Why not? There would seem to be nothing to prevent us from heating a plasma's electrons and atomic fragments to faster and faster speeds, except for one thing. There happens to be a speed limit in the universe: the speed of light in a vacuum, which is 671 million miles per hour (1.08 billion kilometers per hour).

Albert Einstein told us that the electrons in a plasma—or any object, for that matter—may approach the speed of light, but can never achieve it. He also told us that as a particle goes faster and faster, it gets heavier and heavier. For example, when cruising along at 99 percent of the speed of light, an electron has 7 times its normal mass; at 99.999 percent of the speed of light, it is 223 times heavier than when it's not moving.

There must be an ultimate temperature limit, then, lest the particles in a plasma reach the speed of light and become infinitely heavy. Theoretical considerations peg this temperature at around 140,000,000,000,000,000,000,000,000,000,000 degrees—Fahrenheit or Celsius, take your pick.

The next time someone says to you on a blazing summer day, "Whew! How hot can it get?" tell him.

But don't worry. Global warming still has a long way to go.

If Flames Always Go Upward, Why Do Buildings Burn Down?

How do flames always know which way is up?

Light a match and, while it's burning, twist it into a variety of positions. The flame keeps pointing unerringly upward, regardless of the orientation of its fuel. How, indeed, does it "know"?

You are well aware that hot air rises. (If you want to know why, check out p. 107.) A flame, whatever it is, must therefore be carried upward by the rising current of hot air. And that's all we need to know about why flames go upward.

But a more challenging question is, What is a flame? Is it the rising air itself, glowing from the heat? Nope.

A flame is a region of space in which a chemical reaction is going on: *combustion*—a reaction between the oxygen in the air and a flammable gas.

Did I say gas? Yes. But don't solids and liquids burn with flames also? Yes. (And when will I stop asking questions of myself?)

Wood and coal are solids, and they are indeed flammable; gasoline and kerosene are liquids, and they are indeed flammable. But none of them will actually *burn* until they have been converted into a gas or vapor. It's their vapors that burn, because only vapors can mix into the air intimately enough to rub elbows—rub molecules, that is—with the air's oxygen.

Molecules can't react unless they actually come into contact with one another. The oxygen gas in the air can't penetrate the solid or liquid fuel, so the fuel must vaporize and go out to meet the oxygen. That's why we have to *light* a fire. We have to get the fuel hot enough in at least one small location, so it will vaporize. Once the vapor starts burning, the heat of combustion—the combustion reaction releases heat—keeps vaporizing more and more fuel and keeps the process going until all the fuel is gone. (Provided that there is an inexhaustible supply of oxygen.)

A fuel that is already a vapor, such as the methane in our kitchen gas ranges, has no trouble mixing with the air, so it can be ignited by a mere spark. Propane-burning gas grills and butane-burning cigarette lighters contain those fuels in the liquid form, under pressure. But as soon as they are released, they vaporize into gases and mix into the air, whereupon they can also be ignited easily by a spark.

When we light a candle with a match, the match first has to melt a bit of the wax, the liquefied wax must travel up the wick by capillary attraction, and the match must vaporize some of that liquid. Only then can the wax vapor mix with the air and ignite. Without a wick to conduct the liquefied wax up to where there is a good air supply, a candle won't burn.

But if a flame is simply two invisible gases reacting with each other, how come we can see it? In the case of a candle, the flame is visible because oxygen can't flow in fast enough to react completely with all of the rapidly vaporizing wax. So some wax remains unburned as tiny particles of carbon, glowing yellow from the heat and swept upward by the current of hot air.

As the crowd of glowing carbon particles rises higher, the oxygen nibbles away at its outer edges, burning particles up completely into invisible carbon dioxide gas. The crowd of glowing particles is thus depleted more and more as it rises. That's why a candle flame tapers off toward its upper end.

Global Cooling?

Could we counteract global warming if everybody in the world turned on their air conditioners and refrigerators full blast and left the doors open?

Unfortunately, no, for several reasons.

First of all, the world's supply of air conditioners and refrigerators isn't anywhere near what you might think by looking around your neighborhood. But even if every citizen of the less-developed nations were privileged to enjoy cool bedrooms and frozen pizzas, the amount of available coolth wouldn't amount to an ice cube on a glacier. (Yes, I know there is no such word as coolth. Until now.)

Of course, you expected that answer, but maybe not this one: What you are proposing would actually *heat up* the world.

As you know too well from your electric bills, air-conditioning and refrigeration don't come free, in either money or energy. Someone has to produce the electricity that they use, and the production process itself gives off a lot of heat; it's part of the overall energy equation, and therefore part of the environmental problem.

In most cases, the first step in electricity production is to make heat through the burning of coal or by nuclear fission. Then the heat is used to boil water to make high-pressure steam, the steam is used to turn the blades of a turbine and the rotating turbine shaft drives an electricity generator.

That is a remarkably inefficient chain of events, and there's the rub. Or one of the rubs. Only about one-third of the fuel's inherent energy ever winds up as usable electricity. The other two-thirds goes up the smokestack as hot gases or down the river as hot cooling water, or else is lost while the electricity is being transmitted through the wires to your house, because power lines are slightly warmed by their resistance to the electricity flow. That's why birds perch there in cold weather (see p. 100).

More than anything else, then, what power plants really do is heat up the countryside. The more electricity you demand for cooling your food and brood, the more heat the power companies must fling into the environment. Instead of thinking about opening the door of your refrigerator, you'd be doing the world a favor by turning the appliance off!

Okay, you say, but all of that wasted heat is already part of the global warming picture. Turning our refrigerators and air conditioners loose upon the outdoors would have an effect over and above that, wouldn't it?

Again, unfortunately, no.

Consider how a refrigerator or air conditioner does its job. It takes in warm air, removes heat from it and discharges that heat somewhere else. The refrigerator removes heat from the air inside the box and throws it out into the kitchen via coils located behind or beneath the box, while the air conditioner takes in air from the room, extracts heat from it and throws it out the window. But—and here's the main reason your scheme won't work—these machines throw off even more heat than what they remove from the air. You might say that *refrigerators and air conditioners make more heat than coolth.* Here's why.

We know that the natural direction for heat to flow is "downhill" from a higher temperature to a lower one. In order to reverse that natural tendency and force heat to go "uphill" from a cool interior to a warmer exterior, the fridge or AC has to use electrical energy. (That's why you have to plug it in.) And that electrical energy, after it's done its job, turns into heat. You can feel it by touching the outside of the refrigerator or air conditioner; it's warm.

When you add it all up, then, there is more heat—usually about one-third more—coming out of the "cooling" machine than the amount it removed from the box or the room. The bottom line on the energy balance sheet tells us that these machines are actually heating devices.

The final nail in your cool-the-world coffin is this: Even if fridges and ACs could operate without using any electric power, the best you could hope for would be to break even: one calorie of heat discharged somewhere for every calorie removed from somewhere else. And that wouldn't change the world's overall quota of heat. All you'd be doing is moving it around.

NITPICKER'S CORNER

Converting coal or nuclear energy into electricity is, as we've seen, a very inefficient process that puts a lot of waste heat into the environment. But what if your refrigerator and air conditioner ran on electricity from clean, non-fuel-burning sources such as hydroelectric (water), wind or solar power? While they certainly aren't as wasteful as the processes of burning coal and nuclear fuels, these energy sources still can't be converted to electricity with anywhere near 100 percent efficiency. And the waste energy inevitably shows up as environmental heat.

BAR BET   A refrigerator is really a heating machine.

Shocking!

How high does "high voltage" have to be before it's a serious hazard?

Voltage in itself isn't dangerous. A 10,000–volt shock can be no more disturbing than a pinprick, but you can get a serious jolt from a 12–volt automobile battery. What's dangerous is the amount of electric *current* that flows through your body as a result of the voltage.

A current of electricity, as you undoubtedly know, is a flow of electrons. The *voltage* is the amount of push that urges them to flow from one place to another. If they are given no place to flow to, no amount of urging by a voltage will make them flow. Voltage is like height: No matter how high you may be on a cliff, the height is harmless as long as you don't take a direct route to the ground below. Electrical safety is simply a matter of making sure that the electrons can get to the ground by a route other than through your body; they can't hurt you if they're not flowing through you. That's why the birds are safe perching on high-voltage transmission lines (see p. 100).

But it's high time we focused on those electrons that I talk about in several places throughout this book.

Electrons are the negatively charged particles that make up the entire bulk of all atoms. Every atom of every substance is essentially a blob of electrons with an incredibly tiny, incredibly heavy, positively charged nucleus buried somewhere in the middle.

The electrons in atoms have certain energies that are characteristic of the type of atom they're in (see pp. 34 and 37). What makes a flow of electricity possible is that many of these electrons are easily detachable from the rest of their atoms and will travel elsewhere under the influence of a voltage shove. In most cases it takes only a few volts to evict at least some of them from their home atoms.

Some electrons are so loose that you can just rub them off. Scuff your shoes across a carpet on a dry day and some electrons will be rubbed off your shoes' atoms onto the carpet. Because your feet are presumably firmly connected to your shoes, your entire body now has a deficit of electrons, while the carpet has a surplus. Normally, all atoms are electrically neutral, because they have just as much positive charge in their nuclei as they have negative charge in their electrons. But now, your body has fewer electrons than your atoms require.

If you now touch an electron conductor such as a metal radiator or water pipe, electrons from the huge supply in the rest of the world—the ground—will eagerly leap to your finger even before it touches the metal, lighting up the intervening air with a crackling blue spark and inspiring you to utter an expletive. Instead of a water pipe you may even touch another person, who is unlikely to be as electron-deficient as you are, and some of his electrons will jump to your finger, eliciting an expletive from him.

But here's the thing: The voltage that urged the electrons to flow into your finger from the water pipe or your shocked friend may have been several thousand volts, but you're not dead because the number of flowing electrons—the amount of current—was much too small to do any harm to your body. After all, your shoe soles aren't exactly electric generators, like the ones down at the power plant that push gazillions of electrons through transmission lines to your house.

At home, where the voltage has been reduced to 120 or 240 volts, if you touch a "live" wire while some other part of you is connected to the ground, the power company will blindly supply as many electrons as can possibly flow through your body—that is, as large a current as can flow through you, given your body's resistance to the flow. And you're a dead duck.

In short, the danger of electricity lies not in how many volts you are subjected to, but in how much electric current flows through your body. The trouble is that we never know what the current can or will be in any given situation, so we must stay away from any voltage above battery levels at *all* times.

You Didn't Ask, but . . .

If it's the current, not the voltage, that can electrocute a person, how much current is necessary to "do the job"?

Electric current is measured in *amperes*. An ampere is a huge unit of electric current, equivalent to 6 billion billion (6 followed by 18 zeros) electrons passing by every second. So you often hear talk of *milliamperes* or *milliamps*—thousandths of amperes. One milliamp passing through your body will cause a mild tingling sensation. Ten to twenty milliamps can cause muscle spasms that may prevent you from letting go of the "hot" object. Two hundred milliamps, or two-tenths of an ampere, make the heart fibrillate (beat uncontrollably) and can be fatal. Larger currents can stop the heart entirely, but they may not always be lethal because the heart can sometimes be restarted to beat normally again.

A typical automobile battery is capable of delivering a hundred amperes or more; it takes that much current to do the job of turning over an engine. The only reason that auto mechanics aren't dropping like flies is their electrical *resistance;* every substance resists the flow of electricity to a certain degree, and the resistance of human bodies is quite high. That's why it takes a substantial voltage to force enough electrons through a person to electrocute him or her. A 12–volt auto battery doesn't have that much force.

We may encounter dangerous electricity in many different circumstances. I'll assume that you are not terribly concerned about being formally electrocuted while seated in a special chair. But what about lightning? The surge of electrons between a cloud and the ground, or between two clouds, is powered by tens of millions of volts, and that can force tens of thousands of milliamps through the air, which ordinarily won't conduct electricity at all. Get in the way, and a lot of those milliamps can go through you.

How do you "get in the way"? By being close to an object that is offering the lightning's current an easy path to the ground. If ever there were a situation in which the expression "path of least resistance" applies, this is it. The lightning's electrons will flow through the best conductors—materials having the least electrical resistance—that they can find. If you offer them an attractive detour through your body, they'll take it.

Of all materials, metals are the best conductors of electricity; they have the lowest electrical resistance. That's because the electrons in metal atoms are very loose and can flow right along as part of the current. So when sudden thunderstorms have come up on the greens, a bag of metal golf clubs has been many a duffer's ticket to that great fairway in the sky.

Because air is such a poor conductor of electricity, the lightning will take almost any other available path rather than plowing through the air for those last several yards to the ground. Trees, with their nice, juicy sap inside, offer lightning a preferred alternative, so taking shelter from a thunderstorm under a tree may also earn you a trip to the ultimate nineteenth hole. But even if you're out on the seventh green with no trees nearby when a storm comes up, little old you, sticking up only six feet off the ground, can be the lightning's preferred route. Your best lie, so to speak, is flat on the ground, away from your clubs and cart.

Why Doesn't It Rain Roasted Sparrows?

Why don't birds get electrocuted when perching on high-voltage power lines?

This question is as old as electric power itself. It has been asked almost as often as "Do you love me?" and with equally unconvincing replies.

The common answer—"The birds aren't electrocuted because they're not grounded"—doesn't get to the root of the question. Does everyone who walks away after that explanation really know what "grounded" means? What's so special about touching the ground?

As you know, an electric current is a flow of electrons. The key word here is "flow." Unless the electrons can flow from one place to another, they can't do anything useful, or harmful, any more than a stream can turn a waterwheel by standing still. To get electric light, for example, we make electrons flow through a thin tungsten filament, in one end and out the other. In forcing their way through the very thin tungsten wire under the influence of a 115–volt push, they heat it so much that it glows white hot (see p. 50).

Notice that the voltage is the push; that's what voltage is: a force that pushes electrons from one place to another so they can do work for us. But no matter how high the voltage, the electrons can't do anything unless they are given a path to traverse. The power transmission wires are that path. Under the influence of a high-voltage push, they conduct electrons all the way from the power plant to our houses, where they may be tapped off to flow through a lightbulb, a toaster or a television set.

Where do the electrons go after they pass through our electric appliances? They return to Mother Earth, which is where the electric company got them from in the first place. Where else, for heaven's sake, could they have gotten them? The moon? So Mother Earth, whom we familiarly refer to as "the ground," is the original source of electrons at the power company and their final destination when we're done making them work for us. Earth is made of gazillions of atoms containing multigazillions of electrons. By rough estimate, the number of electrons on Earth is 1, followed by 51 zeros. That's what I'd call an inexhaustible supply.

Now, back to the birds. Their little feet are certainly in contact with lots of electrons that are waiting to be drained off and returned to the ground via your electric toaster. But fortunately for the birds, their bodies offer no way of leading the electrons to the ground. The birds just aren't connected to anything; they're a blind alley, an electron dead end. The electrons thus have no way of using the birds as a conduit to the ground, and no electricity flows through them. That's why we don't experience a rain of electrocuted sparrows.

And by the way, what are those birds doing on the power lines in the first place, besides befouling your automobile? In the winter, at least, they are there because the electric current going through the wires generates a small amount of heat that keeps their tootsies warm. And while we're at it, how can they sleep there without falling off? When their foot muscles are relaxed, they tighten up, rather than loosen like ours do. So never fall asleep while hanging from a tree branch.

You may have seen an electric company lineman, raised from a truck in a "bucket," working on electric wires with his bare hands. He's as safe as the birds, because the bucket is completely isolated—insulated—from the ground. Electrons can't find a path through the lineman's body to the ground, so they can't make him glow like a white-hot tungsten filament.

4

The Earth Beneath Our Feet

There are other planets in the universe, but we have a firm attachment to our very own Mother Earth. It's called gravity.[2]

Gravity not only limits our golf drives and makes our body parts sag with age, but serves a number of useful functions, not the least of which is keeping the atmosphere from flying off our spinning Earth like spit from a roller coaster.

Gravity makes dust settle and hot air rise. It does innumerable big and little jobs for us, such as keeping the moon up and skirts down. It even allows us to make electricity from water. Gravity is ubiquitous. Even astronauts don't leave home without it.

This chapter will attempt to tell you how this most wide-ranging of all forces—its effects are felt across the breadth of the universe—operates, even though we can't yet explain what makes it tick, or should I say "stick"?

Earth is, of course, spinning at more than 1,000 miles per hour (1,600 kilometers per hour) as it sails around the sun at more than 10,000 miles per hour (16,000 kilometers per hour). And we're not even dizzy (most of us). But in spite of the fact

---

[2]The name of the gravitational force is *gravitation*, not gravity; *gravity* simply means heaviness. But everybody outside of the Physicists' Club calls the force *gravity*, and whenever I feel like it throughout this book, so do I.

that we are totally oblivious to them (and I'll tell you why), these motions have crucial consequences in our daily lives. They affect hurricanes, ocean currents and ocean tides. They affect— no, they *cause*—every day, night and season of our lives.

In examining the Earth beneath our feet, we'll visit the center of the planet, the North and South Poles, Mount Kilimanjaro in Tanzania, a swirling hurricane and a toilet bowl the size of North America.

And finally, lest we overlook the fact that living things constitute a rather important component of our planet, we'll see how we use radiocarbon dating to explore the past lives of plants, animals and humans.

A Matter of Some Gravity

Why does gravity try to attract all things to the center of Earth? Why to the center? Why not to Mecca, or Disney World?

Because the center of the planet is the center of Earth's gravity: its *center of gravity*.

You've heard the expression "center of gravity" before, and now's your chance to understand what it really means. But first, what is gravity, or, more properly, gravitation?

Gravitation is one of the three fundamental forces in Nature. (The other two are the strong nuclear force, which holds atomic nuclei together, and the electroweak force, which drives certain radioactive changes and is responsible for all electric and magnetic effects.) And what is a force? A force is what makes things move, and nobody can define it any better, despite pages and pages of equations.

The gravitational force acts between any two pieces of matter and tries to bring them together. Every particle of matter in the universe is attracting every other particle of matter, simply because gravitational attraction is an inherent property of matter itself. (And nobody knows exactly why.)

But like two people on an ideal date, gravitation isn't a one-way attraction. It's mutual; each body attracts the other. And the more mass a body has—the more particles of matter it contains—the stronger its aggregate attractive force will be. That's why when you jump off a ladder, Earth doesn't fall upward to meet you. Its superior mass wins out, and Mohammed falls toward the mountain, so to speak.

(If you think you don't know what mass is, pay a visit to the Nitpicker's Corner at the end of this section. If you think you do know what mass is—and if you think you do, you do—then read on.) If mass is what attracts other masses by gravitation, then Earth should attract objects toward wherever most of its mass is concentrated—toward someplace within the body of the planet, rather than to someplace on the surface. But still, why the center?

Consider this: Every particle of matter in the body of our planet is attracting, and being attracted by, all the other particles. A particle that's only a few meters deep beneath the surface is being pulled down by a lot more particles than are pulling it upward, because there are a lot more particles below it than above it. It therefore feels a net downward pull. The same thing can be said for all particles that have more stuff below them than above them, and they are therefore all attracted downward. Downward toward where? Toward the one place that has equal amounts of stuff around it in all directions: the center of Earth. Thus, Earth acts as if it has only one point toward which it attracts everything by gravitation: the center of its gravity.

The center of gravity of a bowling ball, then, is at its geometric center. But every object, no matter how complicated its shape, has a center of gravity. It's the one spot that is the center of all its *masses,* which is not necessarily the center of its shape.

Mother Earth isn't a perfect sphere; she's slightly squashed from north to south and, like the rest of us, she bulges somewhat around her equator. Her diameter through the North and South Poles is 26 miles (42 kilometers) shorter than through the equator. We could still find the geometric center of this slightly unspherical shape and call it the center of gravity, except that Earth's mass isn't distributed uniformly throughout, and it's the center of all the mass that counts where gravity is concerned.

For example, if there were a huge mass of lead buried a few hundred miles below France, Earth's center of gravity would be shifted in that direction. An object dropped in North America would fall slightly more toward France than what is now "straight down." Moreover, France would be closer to Earth's center of gravity than it is now and everything would be heavier. The sound of falling soufflés would be deafening.

An almost incredible application of this principle is the mapping of the oceans' floors by the measurement of slight changes in gravity caused by undersea peaks and trenches. Wherever there is a concentration of mass due to an undersea mountain, the gravitational attraction to the water above is stronger, so water molecules tend to gravitate toward that location, as if to a stronger magnet. That makes a slight pileup of water and a bulge in the ocean's surface that, believe it or not, can be detected by an overhead satellite shooting radar beams down at the sea and watching how they reflect back up. Conversely, where there is a deep-sea trench, the water's surface might be depressed by as much as 200 feet (60 meters). In this way, scientists have made detailed maps of the oceans' floors without even getting wet. Geology books can show you astoundingly realistic pictures of the world's ocean bottoms, as if the waters had been parted by a modern-day Moses.

NITPICKER'S CORNER

An object's *mass* is the amount of stuff or "matter" that it contains. Here on Earth, we measure the mass of an object by seeing how strongly Earth's gravitation pulls it down onto a scale. The more mass, the more pull. We call that amount of pull the object's *weight* (see p. 23).

Of course, the scale is measuring the amount of *mutual* attraction between Earth and the object. But because Earth's pull is always the same, we attribute the scale's reading to the object's *own* attractive force: its mass.

So when your bathroom scale shows a higher reading, you can't attribute it to an increase in Earth's mass. It's that burdensome mass of yours. (Don't say that aloud too fast.)

A Lotta Hot Air

Everybody says that heat rises. But for heaven's sake, why?

Why do they say it, or why does it rise?

They say it because they're speaking carelessly. The statement is just a lot of hot air, because *heat* doesn't rise. What they mean to say is that *hot air* rises.

Heat is one of many forms of energy; it is energy in the form of moving molecules. But it's meaningless to say that any form of energy rises, falls or creeps along sideways. True, energy is always going places and doing things; that's its mission. But it isn't partial to any particular direction, except, of course, for gravitational energy, which on Earth shows a distinct preference for down. (But that's only because the center of Earth—(see p. 104)—lies beneath our feet, which we choose to define as "down.")

We spend our lives engulfed in a sea of air, so when we think of something rising we mean that it's rising *through the air.* Only air or other gases can rise through the air; solids and liquids can't because they're just too heavy, or dense.

That last word, "dense," is the key. The *density* of a substance tells how heavy a given volume or bulk of it is. For example, a cubic foot of water weighs 62.4 pounds (a liter of water weighs 1 kilogram), while a cubic foot of room-temperature, sea-level air weighs about an ounce (a liter of it weighs about a gram). In American Techspeak, we would say that the density of water is 62.4 pounds per cubic foot and the density of air is about 1 ounce per cubic foot. Since there are about 1,000 ounces in 62.4 pounds, we could say loosely that water is 1,000 times "as heavy" (strictly speaking, "as dense") as air.

Now everyone in the world except the United States of America and three other great powers (Brunei, Myanmar and Yemen) uses the International System of Measurement (called "SI," for *Système International* in French), which is apprehensively referred to in the United States as the Metric System. In SI units, the densities of water and air are very simple: 1 kilogram per liter for water and 1 gram per liter for air (at sea level).

But that's room-temperature air. Like most other things, air expands when it's heated, because at higher temperatures its molecules are moving faster and require more elbow room, so they spread out, leaving more empty space between them. More empty space means that a cubic foot (or a liter) of the warmer air will weigh less. It is now less dense than it was.

But the 62.4–dollar question is, What makes that warmer, lighter air move upward through the heavier, cooler air?

Well, what does "heavier" mean? It means that gravity is pulling down on the cool air more strongly than on the warm air. (There are more molecules per cubic foot or liter to pull on.) So wherever warm and cool air find themselves next to each other, the cool air will be pulled down past the warm air. The warm air has no alternative but to get out of the way and be displaced upward. Lo! It is risen.

When one of those beautiful hot-air balloons takes off into the blue sky, people gawking upward from the ground may wonder what force is "pushing it up." Now you know that there is no upward force. That bubble of hot air is merely being subjected to a *lesser downward force,* compared with the cooler surrounding air. And that has precisely, exactly, absolutely the same effect.

NITPICKER'S CORNER

When hot air rises through the atmosphere, the very act of rising cools it off somewhat. I know this sounds paradoxical, but don't turn the page quite yet.

When a warm mass of air rises, it, of course, gains altitude. Masses of air can gain altitude, even if they're not warm, perhaps by drifting up against a mountain and being forced to swoop upward along its slope. Whatever the reason for air masses' moving upward, there must always be equal masses of air moving downward to replace them. The result is that there are rising and falling masses of air all over the world.

Let's see what happens to a particular passel of rising air as it gains altitude.

At higher altitudes, the atmosphere is thinner. That's because there's less atmosphere above it, so it's not under as much compression by gravity. (Gravity pulls air down just as it does everything else; air may be light, but it still has weight.) In other words, at higher altitudes there is less pressure from the atmosphere, and that allows our rising passel of air to expand.

But in order to expand, the passel's molecules have to elbow aside the air molecules that are already occupying that space. And that uses up some of the passel's own energy. What kind of energy? The only energy the air has is the constant flitting-around motion of its molecules. So in elbowing aside the other molecules, the expanding air's own molecules will be slowed down. And slower molecules are cooler molecules, because heat itself is nothing more than moving molecules. (The faster its molecules are moving the hotter any stuff is, and the slower they're moving the cooler it is; see p. 79.) Therefore, as our passel of air rises and expands, it gets cooler.

The higher a mass of warm air rises through the thinner and thinner atmosphere, the more it expands and the more it cools. This is one reason that it's colder up on a mountain than down in the valley. (But see p. 112 for the main reason it's colder at higher altitudes.)

You have undoubtedly experienced the automatic cooling of an expanding gas, whether you paid attention to it or not, because there isn't anybody who hasn't used an aerosol spray can for paint, hair spray, deodorant or whatever. Grab the nearest one, and try this:

TRY IT   Point an aerosol spray can in a harmless direction and spray for three or four seconds. Notice that the can gets cold. It contains a compressed gas—usually propane, now that chlorofluorocarbons (CFCs) have been banished because they chew up the ozone layer. When you press the valve to spray the liquid, the gas is allowed to expand and push the liquid out the nozzle. During that expansion, the gas becomes cooler.

You Didn't Ask, but . . .

Is there any way to tell whether a barbecue grill's propane is going to run out in the middle of a cooking session?

It's pretty hard to look inside that steel tank and see how much propane is left before you fire up the grill, isn't it? Not all grills have pressure gauges.

But hardware stores sell an ingenious little indicator that looks like a strip of plastic because it is. You stick it onto the outside of the tank and, by changing color, it shows you exactly where the propane level is inside the tank. It works by detecting the cooling of the propane gas as it flows out through the valve during use.

The propane inside the tank is under pressure, so it is actually mostly in the form of liquid, with some gas above it. (You can hear the liquid sloshing around if you jostle the tank.) While burning your hamburgers, you are tapping off some of the gas, and more liquid evaporates to replace it. This evaporation cools the gas, so you have a layer of cool gas above a layer of warmer liquid.

The strip of plastic contains liquid crystals, which have different optical properties at different temperatures. What it shows you, then, is one color above the liquid's surface, reflecting the temperature of the cool gas, and a different color below the surface, reflecting the temperature of the warmer liquid. The borderline between the colors is where the liquid's surface lies within the tank.

You'll find that the gauge works only while you're bleeding off gas. After you shut down the tank and it warms up, there is no temperature difference inside, and there are no different colors on the gauge.

Where It's Hilly, It's Chilly

How does a mountaintop, even in the tropics, stay covered with snow all year 'round?

Obviously, because it's always colder up there.

But *why* is it always colder up in the mountains than down at the seashore? After all, doesn't hot air rise (see p. 107)? Shouldn't it therefore be *hotter* up there? There's certainly plenty of hot air in equatorial Tanzania, but Kilimanjaro, which thrusts its peak 19,340 feet (5,895 meters) into the tropical atmosphere, is always capped with snow.

It all starts with the sun. And what doesn't? With the sole exception of nuclear energy, the sun is the source of all the heat and all other forms of energy on Earth (see p. 133).

As the sun shines down on Earth, its light passes quite transparently through the atmosphere, as you must have concluded from the fact that you can see the sun. Not much happens to the light until it strikes the planet's surface. Then, the various types of surfaces—oceans, forests, deserts, car roofs, George Hamilton—absorb the sunlight and are warmed (and in some cases tanned) by it. This makes the entire surface of Earth a giant, warm radiator, and anything nearby—such as the air above it—will also be warmed, just as you are warmed when you stand near the radiator in an old house. (A *radiator,* not surprisingly, is something that *radiates* heat *radiation.* See p. 113.)

It stands to reason, then, that the closer you are to the heat-radiating surface of Earth, the more heat you will be getting from it, just as if you were standing closer to a house's radiator. So the air nearest Earth's surface is warmed the most, and the higher you go away from it, the colder the air will be—cold enough above about 10,000 feet (3,000 meters) that all precipitation will be in the form of snow and it will almost never melt.

(A lesser reason why it's cold in the mountains is that as air masses sweep up the mountainside, they expand because of the lower atmospheric pressure, and when gases expand they get cooler. See p. 94.)

Exactly how does Earth's surface, once warmed by the sun, transmit its heat to the air above it? Mostly by *radiation*—the same way a radiator warms you. But radiation isn't the only way that heat can be transmitted from a warm substance to a cooler one. It can also move by *conduction* and by *convection.* Let's take a quick look at each mechanism.

Conduction: When you grab a hot frying pan handle (DO NOT TRY THIS AT HOME!), the heat travels into your hand by conduction. The heat energy is being conducted, or transmitted, by direct molecule-to-molecule contact. Hot frying pan molecules knock up against your skin molecules and pass their heat energy directly to them. Yelping and releasing your grip breaks this molecule-to-molecule contact. (Actually, the yelp doesn't accomplish much.) Unfortunately, the heat will already be in your skin, continuing to do its damage and replacing your yelp with a more leisurely string of expletives.

(Tip: That heat will stay in your skin, continuing to hurt for a much longer time than you might expect, because flesh is a poor conductor of heat. For a minor burn, get that heat out as quickly as possible by holding it under the cold water faucet.)

Convection: When you open your oven door quickly to peek in at your turkey and you feel a blast of hot air on your face, it's the air that is carrying the heat to you. That's convection: heat being carried on the wings of a moving fluid, such as air or water. In this case, the heat is moving by hitchhiking on the air. When hot air rises (see p. 107), the heat is moving upward by convection. So-called convection ovens are ordinary ovens with fans in them that circulate the hot air around, which speeds up the cooking.

Radiation: The next time you're in a blacksmith's shop (okay, so imagine it) notice that you can feel the heat of his red-hot furnace on your face clear across the room. You're not touching anything hot, so it's not conduction. And there's no moving air, so it's not convection. The heat is reaching you by radiation: infrared radiation.

Infrared is a type of electromagnetic radiation, like visible light except that it has a longer wavelength and human eyes can't see it. What's unique about it is that it is of just the right wavelength that most substances can absorb it, "swallowing" its energy and becoming warmed by it (see p. 34). The infrared radiation isn't heat per se, in spite of what many books may tell you; I call it "heat in transit." It is emitted by hot objects and travels through space at the speed of light, but it doesn't actually turn into heat until it strikes some substance and is absorbed by it. Only a substance can be hot, because heat is the movement of molecules, and only substances—not radiations—have molecules.

You Didn't Ask, but . . .

Does the air keep getting colder and colder without limit as we go higher and higher in altitude?

No, but it does keep getting colder—by an average of about 3.6 degrees Fahrenheit for every thousand feet (6.5 degrees Celsius per kilometer)—up to around 33,000 feet (10,000 meters) above sea level. That's just a bit higher than the cruising altitude of large commercial jet aircraft. You may have heard the airliner's captain try to impress you when flying at that altitude by announcing that the temperature outside your flimsy-looking window was something like 40 degrees below zero Fahrenheit (−40 degrees Celsius). Good thing the window is double-pane-insulated plastic.

Above about 33,000 feet (10,000 meters), you're in the *stratosphere,* where the air stops getting colder as you go higher; it stays roughly constant at about −55 degrees Fahrenheit (−48 degrees Celsius) for the next 12 miles (20 kilometers) or so, and then starts getting *warmer.* Above the stratosphere, the temperature does a couple of other flip-flops, getting first colder and then warmer again.

What's going on?

For one thing, the air has somewhat different chemical compositions at different altitudes. The heavier molecules (carbon dioxide, argon) tend to settle out toward the bottom of the atmosphere, while the lighter ones (helium, neon) tend to rise to the top of the pile. Because those different molecules absorb the sun's energies in different ways, they may heat up differently. The stratosphere, for example, is where most of the ozone molecules live. Ozone absorbs a lot of the sun's *ultraviolet* (very short-wave) radiation, which heats it up and makes the stratosphere warmer than it otherwise would be. Earth's atmosphere is really quite a complicated system.

Beyond the atmosphere? You've heard that the temperature in outer space is extremely cold, haven't you? Well, it isn't. (See p. 175.)

It's Not the Cold, It's the Humidity

I've often heard people say that it's too cold to snow. Is there ever any truth to that?

It's true that when it's very cold it won't snow, but the statement is misleading. Once the temperature gets below freezing and other conditions are right for snow, the will-it-or-won't-it question is purely a matter of the availability of water vapor.

In most cases, in order for it to snow there must first be tiny droplets of liquid water in the air that can freeze into snowflakes. But when the accessible supply of water is very cold, it strongly prefers to stay where it is, namely in the liquid form (see p. 208), so it doesn't contribute much water vapor to the air. Thus, at very low temperatures there just isn't enough water in the air to form those tiny droplets that could freeze and fall as snow.

Of course, if it has been very cold for some time, most of the local water supplies will be inaccessible for vapor production anyway because they're frozen.

In those *National Geographic* pictures of blinding, whiteout blizzards in the Antarctic, it's not snowing—it's blowing. Very strong winds are blowing around loose, already-fallen snow. And when did that already-fallen snow fall? During periods of milder temperatures (but still below freezing), when water vapor was more abundant.

You Didn't Ask, but . . .

Which pole is colder, the North or the South?

The South Pole, where the average temperature is about 56 degrees below zero degrees Fahrenheit (−49 degrees Celsius). At the North Pole the average temperature is a relatively balmy 20 degrees below zero (−29 degrees Celsius).

Antarctica is actually a continent, with the ice and snow lying on top of a huge land mass, whereas the small Arctic ice pack floats atop the Arctic Ocean. The South Pole itself is at an elevation of some 12,000 feet (3,700 meters), and it's always colder at higher altitudes (see p. 112). Moreover, the much bigger ice and snow surface at the South Pole radiates heat away more quickly as soon as the sun goes down.

Yet another factor is that water doesn't get heated or cooled as easily as land does, so it keeps the temperature at the North Pole from going to extremes.

BAR BET   It's much warmer at the North Pole than at the South Pole.

Wheeeeee!

If the whole Earth is spinning at 1,000 miles per hour (1,600 kilometers per hour), why don't we get dizzy, feel the wind or somehow notice the motion? Is it just because we're used to it?

No, it's because Earth's rotation is a uniform, unvarying motion, and we can feel only *changes* in motion (Techspeak: *acceleration*). Any time a moving object is diverted from its motion, either by a change in its direction or a change in its speed, we say that it has experienced an acceleration. Acceleration doesn't just mean going faster.

Say you're a passenger in a car that's moving in a straight line and is operating on cruise control—that automatic speed governor that keeps the car moving at a constant speed. You don't feel any forces pushing your body around, do you? But as soon as the road changes from straight to curved your body becomes aware of it, because you are thrust slightly toward the outside of the curve. Or if the driver suddenly steps on the gas (the "accelerator"), your body becomes aware of it because you are thrust against the back of the seat. Or if the driver suddenly hits the brakes (another accelerator, but a slowing-down one instead of a speeding-up one), your body becomes aware of it because you are thrust slightly toward the front of the car. But as long as the car doesn't speed up or slow down or go around a curve (Techspeak: *angular acceleration*), your body feels no forces trying to push it around. In effect, your body doesn't know it's moving, even if your brain does.

Well, your brain knows that Earth is spinning (see p. 119), but your body doesn't because the motion is smooth, uniform and continuous. As Isaac Newton put it in his First Law of Motion, a body (including yours) that is moving at a constant speed in a straight line will continue moving that way unless some outside force acts on it. Without such an outside force, the body (including yours) doesn't even realize it's moving.

But, you protest, we're certainly being carried around a curve, aren't we? We're following the curvature of Earth's surface. It may be a constant speed, but it isn't a straight line. So why aren't we being thrust outward? Well, we are. But the curvature is so gradual—Earth is so big—that the circular path is *virtually* a straight line, so that the outward force is minuscule (but see p. 122). When you think about it, even your car on that perfectly straight road was going around the same big curve: the curvature of Earth. If you continued in that "straight line" long enough, you'd get right back to where you started.

This is all very discouraging to the diabolical designers of amusement parks (I call them abusement parks), who want us to experience motion to the max. They deliberately make us feel unbalanced, unstable, precarious, disoriented, pushed around and insecure. That's why nothing in the whole place moves at a constant speed in a single direction, except perhaps the outward flow of money from your wallet. Every ride either spins you around, hurls you first up and then down or slings you through some crazy combination of up, down and around at the same time. The best (?) roller coasters are those that combine ups and downs with speedups, slowdowns, twists and curves. These changes of motion, which we certainly *can* feel, all fall into the category of accelerations. Even the merry-go-round is accelerating you, because it is continually diverting you from a straight line, forcing you to turn in a circle.

Oh, you asked why we don't feel the wind as the cosmic merry-go-round named Earth spins us around? It's because the air is being carried around at the same 1,000-mile-an-hour (1,600-kilometer-per-hour) speed as ourselves, our cars, our houses and even our airplanes (see p. 119). So there is no relative motion between us and the air.

This Dizzy World

If Earth is rotating at around 1,000 miles per hour (1,600 kilometers per hour), why can't I see it moving beneath me when I'm in an airplane that's going a lot slower?

Because even when you're flying off to a remote island to get away from it all, you can't escape being part of "it all." Your airplane is attached to Earth almost as tightly as the mountains below are, except that the airplane is (we hope) at a higher altitude.

Your pilot would be the first to assure you that the plane is firmly attached to the air (see p. 22). And since the air is attached to Earth, you might say that we're all in the same boat, sailing merrily eastward along with the surface of Earth at around 1,000 miles per hour (1,600 kilometers per hour).

(The ground's speed is actually 1,040 miles per hour [1,670 kilometers per hour] at the equator; that's the circumference of 24,900 miles [40,100 kilometers] divided by 24 hours. But it's slower as we go north or south on the globe because the circular paths get smaller.)

You do, of course, see the ground "moving" beneath you as you fly. But it's your own airplane's motion that you're seeing, not the ground's. It's the same as seeing the trees "speed backward" as you speed along the highway in your car. That's a very important point to realize: *There is no such thing as absolute motion.* All motion is relative. Nothing can be said to be moving or not moving without specifying "relative to what?" Motion is motion only when it is compared with some independent reference point (Techspeak: a *frame of reference*).

To the trees, you and your car are moving, but to you and your car, the trees are moving. Who's right? If you had been born in your car a second ago, you'd swear that it was the trees that were moving, intuitively and egotistically using yourself as the reference point. It is only with experience that we learn to accept reference points outside of ourselves. If each driver took himself or herself as the reference point, the trees would be "moving" every which way at all kinds of speeds, because every self-centered person's reference point would be moving in a different direction at a different speed. Stationary trees, however, are so much easier to deal with, so we humans have agreed to take the trees and the land they're attached to as our stationary references.

But let's stand back and take a bigger view of Earth. When we say that a palm tree on the equator is moving along with the ground at 1,040 miles per hour (1,670 kilometers per hour), we have to ask, "Relative to what?" Well, how about relative to the center of Earth? That's the only point on or inside the whole globe that isn't moving around in circles. In other words, we're taking the center of Earth as our "stationary" reference point.

But whoa! Let's stand back even farther. The whole planet is moving around the sun at 10,600 miles per hour (17,100 kilometers per hour) relative to the center of the sun, which we can now take as our new reference point.

But the sun itself is moving relative to other stars. And the stars are moving relative to the center of our galaxy. And our galaxy . . .

And on and on, literally ad infinitum.

Before we get too dizzy, let's get back into the airplane. Sitting there, anyplace on the plane is your assumed reference point, so you see Earth "moving backward" with the (forward) ground speed of the plane. But remember that you and your little bag of peanuts and that screaming baby across the aisle are all moving together at approximately Earth's rotational ground speed, relative to the center of the Earth. I say "approximately" because if you're flying eastward in the same direction as Earth's rotation, the plane's speed (relative to the center of Earth) is *added* to Earth's rotational speed; if you're flying westward in the opposite direction of Earth's rotation, the plane's speed is *subtracted* from Earth's rotational speed. If you're flying northeast or south by southwest, consult your high school trigonometry teacher. Can you say "vector"?

NITPICKER'S CORNER

I said that the plane is firmly attached to the rotating Earth because it is firmly attached to the air and the air, in turn, is firmly attached to Earth. Well, not exactly.

Air is a fluid, meaning that it isn't rigid; it flows. So as Earth turns, the air can't precisely keep up; it drags and slops around a bit (see p. 136) like a puddle in a rowboat. Although the plane is indeed firmly held by the air, the air is somewhat loosely held by Earth. That's not to say that we're in any danger of losing our atmosphere; gravity holds that whole layer of air down quite firmly. But within that layer, the air is a churning, blowing, moving mass, and local irregularities can still kick your airplane around with tail winds, head winds and coffee-splattering bumps that make you feel as if you're not very well attached to anything.

You Didn't Ask, but . . .

If I can't see Earth turning from an airplane, can the astronauts see it turning when they look down from their orbiting shuttle?

No, even though they're a lot higher and moving a lot faster than an airplane, the situation is still the same. To them, Earth's surface appears to be "moving backward" at *their* speed of 18,000 miles per hour (29,000 kilometers per hour), just as it appears to you when you're in a car or an airplane. The only difference is that because of their higher speed, they can see a whole continent "moving by" in less time than it probably takes you to drive to work. You may have seen motion pictures of space-walking astronauts with the continents "moving" westward in the background.

But why westward?

Aha! That's an interesting story.

Have you ever wondered why the Kennedy Space Center was built on the east coast of Florida, rather than on the west coast of California? After all, Mickey Mouse is equally accessible on both coasts.

First of all, we want to shoot our rockets out over an ocean, rather than over any populated areas, so that booster rockets can be safely jettisoned. But second and more important, we have to launch our shuttles and satellites into their orbits around the globe by shooting them eastward, in the same direction Earth's surface is moving. That way, we get a free, 1,000-mile-per-hour (1,600-kilometer-per-hour) shove from Mother Earth. And that means the eastward Atlantic Ocean rather than the westward Pacific.

After the shuttle is in orbit, it continues to fly eastward, and looking down, the astronauts see Earth's surface apparently moving westward, just as if they were in an airplane flying from Los Angeles to New York.

But with a lot more legroom.

How to Lose Weight

I'm not sure if this is science or a riddle, but my ten-year-old daughter asked me if a polar bear would weigh less at the equator than it does at the South Pole.

It's both. The riddle part is that polar bears live at the North Pole, not the South Pole. Furthermore, a polar bear at the equator wouldn't be a polar bear, would he? He'd be an equatorial bear.

But let's take the question at face value; namely, will a bear—or anything else, for that matter—weigh less at the equator than it does at either pole? All other things being equal (and, of course, they never are), the answer would be yes. Slightly.

First of all, because Earth bulges out somewhat around the

equator, the bear will be a bit farther from the center of Earth and gravity's pull will therefore be a bit weaker (see p. 104).

But what your daughter undoubtedly had in mind was the effect of our planet's rotation, which is one complete turn every twenty-four hours. (Isn't that a neat coincidence? Of course not. That's how we humans defined twenty-four hours in the first place.) At the equator, which is 24,900 miles (40,070 kilometers) around, that works out to a ground speed—palm trees, bears and all—of 1,040 miles per hour (1,670 kilometers per hour). Back home at the exact North Pole, however, the bear wasn't traveling around at all; he was just rotating in place, at the center of the merry-go-round.

Because of the Earth's rapid rotational speed, bears (and everything else) are subjected to an outward centrifugal force tending to fling them off the planet, just as a dog flings water off his back by rotating himself rapidly after a bath. But the reason that the space around Earth isn't filled with flying bears is that the planet's much stronger gravitational force holds them firmly to the ground.

Nevertheless, the outward-flinging centrifugal force detracts slightly from the Earth-holding gravitational force, so that an equatorial bear's weight is slightly diminished—by a little more than three-tenths of a percent. An 800-pound (360-kilogram) bear would weigh about 3 pounds (1.4 kilograms) less at the equator than he does at the North Pole. In human terms, a 150-pound (68-kilogram) person would weigh half a pound (200 grams) less at the equator than at the North Pole.

Of course, these are the extremes. Anywhere in between the equator and the poles, the rotational speed of the planet is somewhere between zero and the equatorial speed, because the distance around is shorter. So there is a gradual loss of weight as one moves toward the equator from anywhere in the northern or southern hemisphere. If you weigh 150 pounds (68 kilograms) at the latitude of Washington, D.C., and Madrid, Spain, for example, you'd weigh about 5 ounces (14 grams) less at the equator.

That's not a very effective weight-loss strategy, however, unless you get there by walking.

You Didn't Ask, but . . .

Would I weigh less at the bottom of a deep mine shaft than I do on the surface?

Boy, you must really want to lose weight!

Yes, you'd weigh very slightly less.

You're probably thinking that your weight is a consequence of Earth's gravitational pull on your body, and that if you're below the surface gravity has already done part of its job, so there is slightly less pull left. Well, there is something to that, although I'd put it differently.

The gravitational force between two objects acts as if it were coming from the *centers of gravity* of the objects (see p. 104). That is, it acts as if all the mass of each body were concentrated at those precise spots. For a uniform, regularly shaped object such as a sphere, the center of gravity is the same as its geometric center. While Earth isn't a perfect sphere, it is close enough that we can say its gravitational attraction is pulling you toward the center of Earth.

Now when you're on the surface, you are (on the average) 3,960 miles (6,371 kilometers) from Earth's center, the seeming location of all its pull. At the bottom of a mile-deep shaft, you are being pulled toward the center by less Earth-mass than before, because some of Earth's mass is above you and is no longer contributing to the center-directed pull. (Actually, it's pulling you upward.) If there is less mass pulling you to the center of Earth, your weight is less, because that's the definition of your weight: the strength of Earth's center-directed pull on your body.

How much less would you weigh? At the bottom of a ten-mile shaft, you'd weigh about seven-tenths of 1 percent less than at the surface. Not counting all the weight you'd lose digging.

Oddly enough, the *higher* you go above the surface the less you weigh also. You would weigh less on top of a mountain than down in the valley, because you're farther from the center of the Earth.

But wait! Put down that mountain-climbing gear that you bought from those mail-order weight-loss hucksters. If you weigh 150 pounds (68 kilograms) at sea level, you'd weigh only 7 ounces (200 grams) less on top of Mount Everest, the highest point on Earth. Hardly worth the climb, is it? Except for the exercise, of course.

Flushed with Knowledge

Do toilets really flush counterclockwise in the northern hemisphere and clockwise in the southern hemisphere?

No. It's just another one of those urban legends, probably started by an overenthusiastic physics teacher. But it's based upon a grain of truth.

Moving fluids such as air and water are slightly affected by Earth's rotation. The phenomenon is called the Coriolis effect, after the French mathematician Gustave Gaspard Coriolis (1792–1843), who first realized that a moving fluid on the surface of a rotating sphere (Earth, for example) would be deflected somewhat from its path.

And by the way, it's the Coriolis *effect*, not the Coriolis *force*, as so many books and even some encyclopedias refer to it. A force is something in Nature that can move things, such as the gravitational force or a magnetic force. But the Coriolis effect doesn't move anything; it is purely a result of two existing motions—the motion of air or water, as modified by the motion of Planet Earth.

The Coriolis effect is so weak, however, that it shows up

only in huge masses of liquids and gases such as Earth's oceans and atmosphere, where it affects winds and currents quite significantly (see p. 133).

But even if it were much stronger, the Coriolis effect wouldn't show up in a toilet bowl anyway, because the water swirls around for a very different reason: water jets beneath the rim. The toilet designers shoot the water in on a tangent, so as to start it swirling. Of the two toilets in my house, one shoots the water clockwise, while the other shoots it counterclockwise. And they're in the same hemisphere. (It's a small house.)

On the other hand, there are no jets in a sink or bathtub, so when water goes down the drain the direction of its swirl is up for grabs. Draining water must eventually make a whirlpool that turns in one direction or the other, because as its outer portions move inward toward the drain opening, they can't all rush straight to its center at the same time. A whirlpool is the water's way of lining up and taking turns, while still leaving a hole in the middle for the air to come up out of the pipes. If the air didn't have any space for rising to the surface, it would block the water from going down.

But is there any hemispherical preference, no matter how small, for the direction of the swirl in a sink or bathtub?

TRY IT Fill your bathroom sink or bathtub and let the water quiet down for about a week so that there are no currents or temperature differences that could possibly favor one direction over another. Now open the drain without disturbing the water in the slightest. (Good luck.) The water will begin to drain and will eventually form a whirlpool. Repeat this experiment a thousand times and record the number of times it goes clockwise and counterclockwise.

You don't have the time or patience to do this? Good. Forget it. Your sink and bathtub are doomed to failure anyway, because the drain isn't in the center and the water cur-

rents wouldn't be symmetrical. Whirlpools are supposed to be circular.

Scientists who apparently had little else to do have performed this experiment with the biggest, most carefully constructed, temperature-controlled, vibration-free, automatic-central-drain-opening "bathtub" you can imagine, and have been unable to detect any consistent preference for one direction or the other. In other words, it wasn't the Coriolis effect that determined the direction, but various other uncontrollable factors. That's hardly surprising, though, because we can calculate the magnitude of the Coriolis effect to be expected. In a normal-sized bathtub it would be so weak that at most it could push the water around to produce about 1 revolution per day—nowhere enough to overcome the effects of inadvertently caused currents.

Here's the nitty-gritty on how the Coriolis effect works.

Picture Earth as a globe, with North America facing you. Now replace North America with a giant toilet bowl. Its drain opening will be centered somewhere in South Dakota. (No offense, Dakotans.) And let's say that it has no water jets, so that its flushing direction can be determined entirely by Monsieur Coriolis.

The globe, toilet and all, is rotating from your left to your right—from west to east; that's the way Earth turns. But Earth's surface is moving faster at the equator than it is farther north, just as a horsie at the rim of a merry-go-round is going faster than one near the center. That's because a point on the equator has much farther to travel during each rotation than a point near the North Pole does.

Thus, when you drive your car northward from anywhere in the northern hemisphere, the farther north you go, the more slowly the surface of Earth is carrying you eastward. You don't notice this, of course, because you and your car are firmly attached to the surface of Earth and are moving along with it. Air and water, however, are different; they're only loosely attached to Earth's surface, and are free to slop around somewhat. That's why the Coriolis effect can affect only air and water.

Now suppose that you are in the North American toilet bowl, floating in a rowboat somewhere south of the drain opening—say, in Texas. As you start rowing northward toward the drain (away from the equator), Earth under you is carrying you eastward more and more slowly. But your Texan inertia keeps you moving eastward at the faster Texas speed; you are outrunning Earth's surface and getting slightly ahead of it. Net effect? Relative to Earth's surface, you have edged *eastward*. You have been forced into veering slightly to your right, from northbound to slightly eastbound.

Similarly (prove it to yourself), a boat floating southward from Canada would also be deflected to its right: slightly westward. So no matter which direction the water (and your boat) starts out in on its way to the drain, if it's in the northern hemisphere it will always be coaxed into veering to the right. And right turns go clockwise. (But don't go away before visiting the Nitpicker's Corner.)

I'll spare you several more paragraphs of toilet mechanics, but let me just say that in the southern hemisphere everything works the opposite way. Large bodies of moving air and water receive a leftward twist, and therefore tend to swirl counterclockwise. But remember: The body of water has to be huge before you can see much effect. Oceans, yes; toilets and bathtubs, no.

NITPICKER'S CORNER

Okay, so tornadoes and hurricanes really rotate *counterclockwise* in the northern hemisphere and *clockwise* in the southern hemisphere—exactly the opposite of what I just led you to believe. Just hold your horses and everything will turn out right. Or left. Whatever. Lemme 'splain it to ya.

And let's stay in the northern hemisphere, okay?

Hurricanes form in areas of low air pressure. That means that the air there is distinguishably less dense, less heavy than the air surrounding it; it's sort of like a hole in the air. Now if, because of the Coriolis effect, all the heavier air surrounding the "hole" is glancing off it to the right, that would make the "hole" itself rotate to the left. Thus, the resulting low-pressure hurricane spins counterclockwise.

No? Well, how about this? The low-pressure zone is a roulette wheel and you are the higher-pressure air. While thrusting your hand to the right, you brush it against the wheel's edge. Won't that make the wheel spin to the left?

Or this: You're pushing some kids around on one of those little playground merry-go-rounds, carousels, roundabouts, whirligigs or whatever they're called. You push it to the right and the kids spin to the left. Right?

Or—oh, hell. Just look at the diagram on page 130.

And what about the southern hemisphere? Just interchange all the "lefts" and "rights" in the last four paragraphs and all the "clockwises" will run the other way.

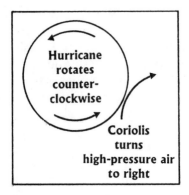

BONUS: Here is your reward for reading all of the foregoing without your head spinning either clockwise or counterclockwise: I'm going to tell you why all our clocks run clockwise.

It's because the first mechanical clocks were invented in the northern hemisphere.

Not obvious? Consider this.

To an observer in the northern hemisphere, the sun is always somewhere in the southern sky. Looking southward toward the sun, a northern-hemisphere observer sees it moving across the sky from east to west, which to him is from left to right. The hour hands on early clocks—and at first there were *only* hour hands—were intended to mimic this left-to-right movement of the sun. Hence, they were made to move across the top of the dial in the direction that we now call "clockwise." When the refinement of minute hands came along toward the end of the sixteenth century, they, of course, were made to go in the same direction. Can you imagine a clock with the hour hand going one way and the minute hand going the other?

BAR BET If mechanical clocks had been invented in Australia, they'd all be running counterclockwise.

The Infernal Equinox

Is it true, as some people claim, that during the vernal equinox it is possible to stand an egg on end?

Absolutely. And during the autumnal equinox as well. And on Tuesdays in February, and anytime during the fourth game of the World Series when the count is three and two on a left-handed batter. Get the picture?

The point, of course, is that equinoxes have nothing whatsoever to do with balancing eggs. But old superstitions never die, especially when perpetuated year after year by kooks who like to chant and perform pixie dances in the meadows on the day of the vernal equinox.

You can balance an egg on end anytime you feel like it.

TRY IT   Take a close look at an egg. It isn't glassy smooth, is it? It has little bumps on it. Go through a dozen and you're sure to find several that are quite bumpy on their wide ends.

Now find a tabletop or some such surface that is relatively smooth, but not glassy smooth. With a steady hand and a bit of patience, you'll be able to accomplish this miraculous astronomical (more appropriately, astrological) feat without any contribution from Mother Earth, except for supplying the gravity that makes the task challenging. If the balancing surface is rather rough, like a concrete sidewalk, a textured tablecloth or a low-pile rug, for example, it's a piece of cake. An old after-dinner trick—on any day of the year—was to conceal a wedding ring under the tablecloth and, with feigned difficulty, "balance" the egg on it.

So much for the old egg game. But what is an equinox, anyway?

Picture Earth, circling the sun at the rate of 1 revolution per

year. The circle made by Earth's orbit around the sun lies in a plane, just as a circle drawn on paper lies in the plane of the paper; it's called the *plane of the ecliptic.* Now Mother Earth wears another circle around her middle; it's called the equator, and it also lies in a plane, called the *equatorial plane.* We can imagine the equatorial plane being extended beyond Earth, way out toward the sun. Funny thing, though: It misses the sun. You usually won't find the sun anywhere in the equatorial plane. That's because Earth is tilted, so its equatorial plane passes above or below the position of the sun. (The equatorial plane is tilted from the plane of the ecliptic by 23 ½ degrees.)

As the tilted Earth moves around the sun, there will be two times in the year when the two planes intersect—that is, two times when the sun, in its ecliptic plane, is also in the equatorial plane, meaning that it is directly over the equator. For half of the year, the sun is north of the equator and the northern hemisphere has spring and summer; for the other half of the year the sun is south of the equator and the southern hemisphere has spring and summer. The two "crossover" instants usually occur on March 21 and September 23. Those two instants are how we define the beginnings of spring and

fall in the northern hemisphere; they are called the vernal (spring) equinox and the autumnal (autumn) equinox.

The word *equinox* comes from the Latin meaning equal night, because at those instants the periods of daylight (the days) and darkness (the nights) are of equal duration all over the world. You can see that from the fact that the sun is directly over the equator, favoring neither more daylight in the north nor more daylight in the south.

Without knowing all of this, primitive people found the equal-light-and-darkness dates to have special significance, ushering in, as they do, seasons of warmth and growth or cold and barrenness. So all sorts of superstitions grew up around these dates. You can see, though, that there is no "alignment of the planets" or any other possible gravitational effects of the equinoxes that would make eggs do anything weird. The only things that are weird are the nuts who still claim that these dates have magical powers.

Oh, yes, then there's the matter of the *solstices*. They occur halfway between the equinoxes. The summer and winter solstices are the instants at which the sun gets as far north or south of the equator as it ever gets during the year. For northern hemispherians, the summer solstice falls on June 21 or 22 and the hours of daylight are longest; you might call it "maximum summer" or "midsummer." It has no more mystical power over eggs than the equinoxes do, although in Scandinavia, where the winters are long and dark and Midsummer Day is an excuse for great revelry, it does seem to have a mysterious effect on alcohol consumption.

O, Solar Mio

When the world runs out of coal and petroleum, could we get all our power from solar energy, which is inexhaustible?

Probably not, if you mean making electricity from solar panels.

There certainly is lots of sunshine, but capturing it and converting it efficiently is the problem. Let's do the arithmetic.

Every day, the sun shines down upon the surface of Earth an amount of energy equal to three times the world's annual energy consumption. That means that to keep up with consumption we would have to capture and convert all the sunlight falling on about one-tenth of a percent of Earth's surface. That may not sound like much, but it's about 180,000 square miles (470,000 square kilometers) of solar panels, or about the size of Spain. Double that to take care of the inescapable fact that it's always nighttime in half the world. And oh, yes: There are clouds.

But if you think about it, all of our energy sources *today* come from the sun, with only one exception: nuclear energy, which we discovered how to make about sixty years ago. Nuclear energy, in the form of nuclear fusion, is where the sun gets its energy in the first place. So speaking cosmically, there is really only one source of energy in the universe, and it's nuclear. Even Earth's internal heat, the source of volcanos and hot springs, is fed by nuclear energy from radioactive minerals.

But until we learned how to make some of our own nuclear energy down here on Earth, we had to procure our share of cosmic nuclear energy through a go-between: Old Sol. The sun converts its own nuclear energy into heat and light for us, and all of our current energy sources come from that heat and light. They are therefore solar energy in a real sense.

Let's look at our "solar energy" sources one at a time.

Fossil fuels: Coal, natural gas and petroleum are the remnants of plants and animals that lived millions of years ago. But what created those plants and animals? The sun. Plants used the energy of sunlight to grow by photosynthesis and the animals came along and ate the plants (and, alas, one another). All life on Earth owes its existence to the sun and so, therefore, does the energy we get today from fossil fuels.

Water power: Hydroelectric power plants suck the gravitational energy out of falling water by enticing it into plummeting down onto the blades of turbines, our modern version of the waterwheel. Instead of your having to have a waterwheel or turbine in the kitchen to grind your coffee beans, the turbine-driven generators convert the water's gravitational energy into electrical energy, which is then piped to your wall outlet through copper wires.

The water cascades down Niagara Falls or spills over Hoover Dam because in deference to Sir Isaac Newton (the falling-apple guy), it is trying to get closer to the center of Earth (see p. 104). Then isn't water power really the power of gravitational attraction? Isn't it Earth-provoked, rather than sun-provoked?

Yes, but hold your horsepower. How did that water get so high in the first place that it can then fall down under the influence of gravity?

It's the sun again. The sun beats down on the oceans, evaporating water into the air, where it is blown around by the winds, forms clouds and eventually rains or snows back down. So without the sun's water-lifting power, we wouldn't have water-falling power. We wouldn't have waterfalls or rivers, because without being replenished from above by sun-raised rain and snow, they'd all run dry.

Wind power: Windmills capture energy from moving air. But what makes the air move? You guessed it: the sun.

The sun's rays shine down upon Earth's surface, a little stronger here and a little weaker there, depending on the seasons, the latitudes, cloud cover and a number of other things. But the land is warmed up by the sun's rays much more than the oceans are, and that creates unevenly heated air masses around the globe. As the warmer masses rise and the cooler masses rush in at ground level to replace them (see p. 107), the air flows, producing everything from balmy breezes to monsoons. Because all of these winds are ultimately traceable back to the sun's heat, wind power is truly sun-provoked.

NITPICKER'S CORNER

All right, so all of our winds aren't caused by the sun. Some of them are caused by Earth, without any outside help.

Earth is rotating, and as it rotates it carries along a thin surface layer of gas—the atmosphere. Now gases and liquids are what we call *fluids,* substances that flow easily, unlike solids. (Most people use the word "fluid" to mean only liquids, but gases also flow, and are therefore fluids.) Any fluid will have a tough time staying in place when the solid body it's trying to hang on to is moving. In an airplane, for example, the coffee in your cup slops around when the plane hits bumpy air the moment after the flight attendant pours it.

In the same way, the rotational movement of Earth makes the air slop around to a certain extent, like the coffee in the cup. And what is air that's slopping around? Wind. So some of our winds are Earth-provoked, rather than sun-provoked.

One way in which Earth's rotation affects air movements is quite interesting. It's called the Coriolis effect (see p. 125).

How to Date a Mummy

Can radiocarbon dating tell us how old <u>anything</u> is?

It won't help you to determine the age of anything that is still alive, such as a twelve-year-old posing as a twenty-five-year-old in an Internet chat room. Radiocarbon dating is useful for determining the ages of plant or animal matter that died anywhere from around five hundred to fifty thousand years ago.

Ever since its invention by University of Chicago chemistry professor Willard F. Libby (1908–1980) in the 1950s (he received a Nobel Prize for it in 1960), the radiocarbon dating technique has been an extremely powerful research tool in archaeology, oceanography and several other branches of science.

In order for radiocarbon dating to tell us how old an object is, the object must contain some organic carbon, meaning carbon that was once part of a living plant or animal. The radiocarbon dating method tells us how long ago it lived, or more precisely (as we'll see), how long ago it died.

Radiocarbon tests can be done on such materials as wood, bone, charcoal from an ancient campfire or even the linen used to wrap a mummy, because linen is made from fibers of the flax plant.

Carbon is the one chemical element that every living thing contains in its assortment of biochemicals—in its proteins, carbohydrates, lipids, hormones, enzymes and so on. In fact, the chemistry of carbon-based chemicals is called "organic chemistry" because it was at one time believed that the *only* place that these chemicals existed was in living organisms. Today, we know that we can make all sorts of carbon-based "organic" chemicals from petroleum without having to get them from plants or animals.

But the carbon in living things does differ in one important way from the carbon in nonliving materials such as coal, petroleum and minerals. "Living" carbon contains a small amount of a certain kind of carbon atom known as carbon-14, whereas "dead" carbon contains only carbon-12 and carbon-13 atoms. The three different kinds of carbon atoms are called *isotopes* of carbon; they all behave the same chemically, but they have slightly different weights, or, properly speaking, different masses (see p. 23).

What's unique about the carbon-14 atoms, besides their mass, is that they are radioactive. That is, they are unstable and tend to disintegrate—break down—by shooting out subatomic particles: so-called beta particles. All living things are therefore slightly radioactive, owing to their content of carbon-14. Yes, including you and me; we're all radioactive. A typical 150-pound person contains a million billion carbon-

14 atoms that are shooting off 200,000 beta particles every minute!

BAR BET   You are radioactive.

If the world's nonliving carbon isn't radioactive, where do living organisms get their carbon-14? And what happens to it when the organisms die? The answers to those questions is where the radiocarbon story really gets exciting. Professor Libby, working right down the hall from my laboratory at the University of Chicago, was able to recognize the relationships among a series of seemingly unconnected natural phenomena that, when put together, gave us an ingenious method for looking into our ancient past and into the history of our entire planet. Follow this sequence of events.

(1) Carbon-14 is continuously being manufactured in the atmosphere by cosmic rays, those high-energy subatomic particles that are shooting through our solar system in all directions at virtually the speed of light. (Some of them come from the sun, but the rest come from outer space.) When these cosmic particles hit Earth's atmosphere, some of them crash into nitrogen atoms, converting them into atoms of carbon-14. The carbon-14 atoms join with oxygen to become carbon dioxide gas, which mixes thoroughly around in the atmosphere because of winds. So the entire atmosphere has a certain amount of carbon-14 in it, in the chemical form of carbon dioxide. This process has been going on for eons, and the carbon-14 in the atmosphere has settled into a fixed amount.

(2) The radioactive carbon dioxide is breathed in by plants on Earth's surface and used to manufacture their own plant chemicals. (You know, of course, that plants take in carbon dioxide to use in photosynthesis.) All plants on Earth

therefore contain carbon-14. They all wind up with about 1 atom of carbon-14 for every 750 billion atoms of carbon that they contain.

(3) For as long as a plant is alive, it continues the process of taking in atmospheric carbon dioxide, thus maintaining its 1-in-750-billion atom ratio of carbon-14.

(4) As soon as the plant dies it stops breathing in carbon dioxide and its accumulation of carbon-14 atoms, no longer being replenished by the atmosphere, begins to diminish by radioactive disintegration. As time goes by, then, there are fewer and fewer carbon-14 atoms remaining in the dead plant material.

(5) We know the exact rate at which a number of carbon-14 atoms will diminish by radioactive disintegration (visit the Nitpicker's Corner). So if we count how many of them are left in some old plant material, we can calculate how much time has gone by since it had its full complement of 1 in 750 billion and, hence, how long ago the plant died. In the case of a piece of wood, for example, we will know when the tree was cut down; in the case of a mummy, we can measure its linen wrapping and calculate when the flax plant was harvested to make the linen, and so on. Neat, huh?

But what about animal relics such as bones and leather? How can we tell when an animal lived and died? Well, animals eat plants. Or else they eat animals that have eaten plants. Or in the case of human animals, both. So the carbon atoms that animals eat and from which they manufacture their own life chemicals have the same ratio of carbon-14 atoms as the plants do: 1 out of every 750 billion. When the animal dies it stops eating and exchanging carbon atoms with its surroundings, and its load of carbon-14 begins to diminish in its precisely known way. By measuring how much carbon-14 remains today, we can calculate how much time has elapsed since the relic was part of a living animal.

There have been several spectacular applications of radiocarbon dating in the past few decades. One of these was the dating of the Dead Sea Scrolls, a collection of some eight hundred manuscripts that were hidden in a series of caves on the coast of the Red Sea, ten miles east of Jerusalem, by Essene Jews around 68 B.C. They were discovered by Bedouin Arabs between 1947 and 1956. The linen-wrapped leather scrolls contain authentic, handwritten portions of the Old Testament that were determined by radiocarbon dating to have been written around 100 B.C.

Another triumph of radiocarbon dating was the finding that the Shroud of Turin, believed by some to be the burial cloth of Jesus, is a medieval fake concocted sometime between 1260 and 1390 A.D., which is very A.D. indeed. This unambiguous scientific result, obtained independently in 1988 by three laboratories in Zurich, Oxford and Arizona, continues to be rejected by those who prefer to believe what they prefer to believe.

NITPICKER'S CORNER

How do we know precisely at what rate the amount of carbon-14 diminishes?

Every radioactive, or unstable, atom has a certain probability of disintegrating within a certain period of time. Some kinds of radioactive atoms are more unstable than others and have higher probabilities of disintegrating. We can't tell when any single atom will disintegrate, but averaged over the zillions of atoms in even a minute speck of radioactive matter, the statistics are completely predictable. It's like tossing coins: You have no idea whether any single toss will be heads or tails, but you know for sure that if you toss the coin a zillion times, there will be half a zillion heads and half a zillion tails.

In the case of radioactive atoms, the statistics are such that one-half of the atoms disintegrate within a certain amount of time called the *half-life*. And that's true no matter how many of those radioactive atoms you start with.

The half-life of carbon-14 has been measured to be 5,730

years. Start out with a zillion carbon-14 atoms, and 5,730 years later there'll be half a zillion of them left. After another 5,730 years, there'll be only a quarter of a zillion remaining, and so on. So if we count the number of carbon-14 atoms in a sample of old wood and find that it contains exactly half as many as in a similar piece of living wood, then we know that it was cut from the tree 5,730 years ago. And so it goes for any amount of carbon-14 and any amount of time, although the math isn't as simple. Calculus and logs and stuff. You don't want to know.

5

Heavens Above!

One difference between humans and animals is that animals never look up at the sky. All their food lies within a thin layer on or near the ground that biologists call the biosphere. And food is all they need.

But our human need for nourishment is spiritual and intellectual as well as physical. From the first moment we began to wonder "how" or "why," we have always looked to the heavens for answers to our wondering.

The heavens—the great *up there*—have always held for us a mystical attraction. The heavens are a conceptual sublimation of everything that lies beyond our comprehension. Earliest man looked *up there* and wondered what the stars were. Then we invented gods, and where else should we establish their home offices but *up there*? Heaven, the ultimate unknown beyond death, could be placed nowhere else.

Later in human history, in an attempt to build a more tangible bridge to the heavens, astrologers concocted an intricate web of supposed associations between the motions of the stars and planets *up there* and all of our motions and emotions down here. Incredibly, in the twenty-first century there are still those who believe that a planet a billion miles away can tuck a winning lottery ticket into their pockets.

Today, having explored everything from the ground on up as far as the outer edges of Earth and beyond, we find much less mystery remaining in the *up there*. We can fly not only to the top of the sky, but beyond it to other planets. We now have to focus on a more distant realm of the unknown, the *out there* of space, a whole universe of unimaginable secrets that will continue to evade us, perhaps forever. We continue to look upward and wonder.

In this chapter we will first explore the lowest level of sky, the atmosphere, which not only sustains all life with its oxygen (for animals) and carbon dioxide (for plants), but conveys all light to our eyes, sound to our ears and scents to our noses. We'll see the moon turn blue, we'll hear a sonic boom emanating from a lion's cage and we'll smell some absolutely disgusting stuff. Then we'll turn out the lights and look at the night sky, which has never ceased to enchant humankind. Do you *really* know why the stars twinkle and the moon doesn't?

And finally, we'll leave Earth and venture into outer space, where it's really, really cold. Or is it?

P-U!

When I'm smelling some really disgusting stuff, are little pieces of that stuff actually entering my nose?

Sorry, but yes. Not actual fragments, but individual molecules—molecules that have evaporated from the "stuff" and have floated through the air to your nose.

But don't get sick at the thought. It takes only an incredibly small number of molecules to be detected by humans as an odor. And the molecules aren't even molecules of "the whole stuff."

Virtually every kind of stuff you can imagine (and others that you may not even care to imagine) are complex mixtures of many different chemical substances. Each of these chemicals has a certain tendency to send some of its molecules off into the air as a vapor. The molecules that enter your nose are not a gross (pun intended) representation of the entire, disgusting stuff, but only the molecules of its most volatile (easily evaporated) chemical components. When you say "I smell stuff X," it's because you have learned to associate the smell of those few volatile chemicals with the entire, chemically complex stuff X. Individually, any particular one of these chemicals in its pure form, removed from its disgusting context, may be quite innocent, even though smelly.

Nevertheless, several unpleasantly odoriferous chemical compounds have been named for the "stuff" that they are found in, and that they are largely responsible for the odor of. Caproic acid is so named because it smells like goats. (*Caper* is Latin for goat.) *Cadaver*ine is a chemical component of putrefying flesh. And skatole smells like . . . well, *skatos* is the Greek word for excrement.

Most astounding fact of the week: Skatole is used in perfumes. Yes, it's a fixative, which keeps perfumes from evaporating too fast—but not from being described in impassioned, romantic terms by advertising copywriters.

If they only knew.

Vacuum Cleaners Suck!

What would happen if I operated a vacuum cleaner in a vacuum?

You'd get an exceedingly clean vacuum.

But seriously, I don't know why you'd want to imagine a thing like that, because there is nothing cleaner than a true vacuum; it is the epitome of nothingness. I'll assume, however, that you ask the question out of scientific curiosity, rather than because it's funny.

What is a vacuum? People use the word very loosely to describe any space that contains something less than its normal complement of atmospheric air molecules. In normal air at sea level, a cubic inch of air contains about 400 billion billion molecules. (That's 27 billion billion molecules per cubic centimeter.) Suck some of them out by any means at your disposal—a sipping straw, a vacuum cleaner or a vacuum pump—and you're allowed to call the space a vacuum. But it's really only a *partial* vacuum; there's still lots of air in it. A vacuum cleaner can't even pump out half the air in a container.

A *perfect* vacuum, a *real* vacuum, on the other hand, is a space that contains absolutely nothing, not even a single molecule. But a perfect vacuum is only an abstract concept, like a perfectly trustworthy politician. It just doesn't exist in the real world.

Why? Because even if you could invent a 100 percent efficient vacuum pump that could suck every last molecule of air out of a container—and you can't, for a reason that will very soon become apparent—the container itself would be sloughing off molecules of itself into the pumped-out space. That's because absolutely every substance in the world has a *vapor pressure* (see p. 208)—a certain tendency for its molecules to fly off into space as vapor. That's true no matter how solid and Gibraltar-like the substance may appear to be. A scientist would say (and I will) that there is an *equilibrium*—a balance—between the substance in the solid form and the same substance in the vapor, or gaseous, form (see p. 200). Every molecule on the surface of a solid has the option of staying attached to the solid or flying off into space as a gas molecule.

All I'm saying about solids is what you already know to be true of liquids: that molecules of a liquid can go flying off into space as a vapor. Water, for example, e*vapo*rates (becomes vapor) at a pretty good clip; its vapor pressure is fairly high. Oils, on the other hand, don't evaporate very much; their vapor-producing tendencies, or vapor pressures, are low.

Much, much, much lower than any liquid are the evaporating tendencies of solids. You've never seen a piece of iron "dry up" and disappear into the air, have you? But that doesn't mean that, now and then, an occasional iron atom isn't breaking its attachment to its solid buddies and sailing off into the wild blue yonder.

To put things in perspective: The tendency of liquid water to evaporate is 500,000,000,000,000,000,000,000,000,000,000, 000,000,000,000,000,000,000,000,000 times higher than that of solid iron. But that still doesn't mean that you could build a *perfect* vacuum chamber out of iron. There'll always be a few iron atoms floating around in it. Moreover, what would you use to seal it up airtight? Rubber gaskets? Rubber has a very significant vapor pressure and there will be lots and lots of rubber molecules in your "vacuum" space.

And so on. Even if you could build a vacuum chamber entirely out of tungsten metal, which has the lowest vapor pressure of any known substance—something like one or two atoms flying around in the entire universe—you still couldn't pump it out completely because *the vacuum pump itself* is made of stuff like gaskets, oil and grease, etc., all with their own significant vapor pressures.

All this hasn't prevented scientists from trying to produce the best possible vacuum. The best they've been able to do so far is a space that contains only a few million molecules per cubic inch or cubic centimeter, as compared with the 27 billion billion molecules in ordinary air. That's emptier than a wallet just before payday.

But your question implied that you wanted to stand in a completely evacuated room (if you could survive there) with a vacuum cleaner in your hand, and you wanted to know what the vacuum cleaner would suck in. Nothing. The fan would just go 'round and 'round without sucking or blowing anything, because there's nothing to suck or blow.

But you knew that, didn't you, you rascal.

The Crack of Boom

When a lion tamer cracks his whip it makes a very loud "crack." But he's not hitting the lion and it looks as if the whip isn't even touching the ground. What makes the loud noise?

The crack of a bullwhip is actually a miniature sonic boom, produced because the tip of the whip is traveling through the air faster than the speed of sound.

When the cat master snaps his whip sharply, he's putting a great deal of energy into the handle end. That energy has no place to go except to travel down the length of the whip as a wave of motion. In Techspeak, energy of motion is called *kinetic energy,* and it's a function of both weight and speed (actually, mass and velocity, but let's not quibble). A given amount of kinetic energy can come from a heavy object moving relatively slowly or a light object moving relatively fast. For example, in order to match the kinetic energy of a ten-ton truck moving at 50 miles per hour (80 kilometers per hour), a one-ton automobile would have to be traveling at 158 miles per hour (254 kilometers per hour).

(The mathematically unchallenged will immediately recognize that those speeds aren't inversely proportional to the weights. That's because kinetic energy is proportional to the *square* of the velocity.)

As the energy moves down the length of a bullwhip it has less and less mass to work with, because the whip is tapered. The energy has to stay within the whip because it has no place else to go, so as the thickness and weight decrease the velocity has to increase.

Have you ever played "crack the whip" on ice skates? A long line of skaters travels in unison, and when the lead skater makes a turn a wave of turning energy accelerates down the line until the last guy is yanked around so fast that he can barely hold on. In a long bullwhip snapped hard, the speed at the tip can easily exceed the speed of sound and create a small sonic boom.

What happens to the energy when it gets to the tip of a whip? If you examine a well-used one, you'll see that many of the "guys at the end" have actually been snapped off; the tip is frayed. But much of the energy has gone directly out into the air as sound, while some of it is reflected back up the length of the whip. The reflection turnaround at the tip is incredibly fast, and that fast-reversing wave also contributes to the noise.

Now all we need to understand is why lion tamers ever decided to use chairs. You'd think they could find something more sophisticated and professional-looking in the Tamers "R" Us store.

You Didn't Ask, but . . .

What causes a sonic boom?

There's a lot of nonsense out there about sonic booms. The *Columbia Encyclopedia* 5th edition (1993) says, "An object such as an airplane generates sound. When the speed of the object reaches or exceeds the speed of sound, the object catches up with its own noise" (I wish some politicians would do that), which causes "piled-up sound." Ridiculous! Will somebody please tell me what a pile of sound is supposed to be?

On the other hand, many people believe that there is a tangible thing called "the sound barrier," and that when an airplane passes through it it makes a crashing sound, as if crashing through an invisible wall of glass. That's wrong too. I guess people have been led to think that way because of the word "barrier." It was never meant to imply that there was a physical obstruction up there in the air, but only that the speed of sound posed an obstruction to the development of faster and faster airplanes. It was an aeronautical design barrier, not a physical one. Nevertheless, when an airplane "crosses" the sound barrier there certainly is a lot of physical stress on the plane because of the shock wave, as we'll see.

The actual barrier to supersonic flight is imposed by the speed of sound itself. (And by the way, *supersonic* means faster than the speed of sound; *ultrasonic* refers to sound of a higher frequency than humans can hear.) Unique things do indeed happen when an object approaches the speed of sound in air. Here's what goes on.

Air, of course, consists of molecules: molecules of nitrogen and oxygen, mainly. In all gases, the molecules are flitting frenetically through space in all directions like a swarm of maniacal bees. At room temperature, for example, the oxygen molecules in the air are zipping around at an average speed of 1,070 miles per hour (1,720 kilometers per hour). The hotter the gas is, the faster the bees are flying (see p. 107).

An airplane flying through the air at a paltry few hundred miles or kilometers per hour gives these sprightly molecules plenty of time to get out of the way and let it through; it's like a person wending his way slowly through a crowd. But when the plane's speed becomes comparable to the molecules' own speed, they don't have time to get out of the way; they just pile up on the front edges of the plane and get pushed along in front of it like snow before a plow. This rapid pileup of compressed air constitutes an "air shock" or shock wave, which is, in effect, a loud noise. The sound waves radiate out in all directions and can be heard as a "boom" on the ground below. The plane carries its "circle of boom" along with it, so that people on the ground along the plane's path will hear it when the plane passes over them. This explains away the popular misconception that there is a single boom as the plane crosses the sound barrier. It is a traveling boom.

What does all that have to do with the speed of sound?

Well, sound is nothing but a series of compressions and expansions in the air (see p. 28). If the air's molecules are flitting around at some particular speed, there will be a limit to how fast that air can be compressed and expanded, because the molecules can't be compressed and expanded any faster than they can advance and retreat to and from one another. Thus, the speed of the air's molecules imposes a limit on how fast they will permit sound to pass through—a limit on the speed of sound through that particular air.

Sound will travel faster in warm air than in cool air (see p. 28), because warmer molecules are moving faster and can collide with one another more effectively (see p. 107). Example: The speed of sound at sea level is 947 miles per hour (1,524 kilometers per hour) at 80 degrees Fahrenheit (27 degrees Celsius), but only 740 miles per hour (1,200 kilometers per hour) at 32 degrees Fahrenheit (0 degrees Celsius). Sound also travels faster in dense high-pressure air because the molecules are closer together and can better transmit compressions.

Putting it all together, then, the speed of sound is fastest in warm, sea-level air and slowest in cold, thin air. That's why supersonic aircraft operate best at frigid high altitudes, where they don't have to go quite so fast to exceed the speed of sound. At 30,000 feet (9 kilometers) above sea level, the air is cold enough and thin enough that the speed of sound is only 680 miles per hour (1,100 kilometers per hour).

On Donner und Blitzen

Why does thunder sometimes sound like a sharp crack, and sometimes like a low rumble?

It depends on how far you are from the lightning. The closer you are, the higher the pitch of the sound you hear; the farther away you are, the lower the rumble.

First, we have to remind ourselves of what thunder *is*.

A stroke of lightning (see p. 97) is extremely fast; it occurs with what might be called lightning speed. Its sudden heat makes the surrounding air white hot—heated to tens of thousands of degrees. The air expands at tremendous speed, after which it rapidly cools and contracts back to its normal temperature and pressure. Air moving so suddenly makes huge vibrations, and that's what sound waves are: shudders, or pressure waves, moving through the air. Hence, the noise of thunder.

It will not surprise you to learn that thunder travels at the speed of sound. But light travels almost a million times as fast as sound. Obviously, then, you're going to see the lightning flash almost instantaneously, but you won't hear the thunder until it travels from the lightning strike to your ears.

TRY IT The next time you have the privilege of witnessing a bang-up thunderstorm, count the number of seconds between a lightning flash and the beginning of the associated thunderclap. Divide that number of seconds by 4 to find out roughly how many miles away the lightning was. Or multiply the number of seconds by 400 to get the approximate distance in yards. (But see the Nitpicker's Corner.) You may be shocked—sorry, I mean surprised—to find how close many of the lightning strikes are. And while you're at it, notice that the closer the lightning is, the higher-pitched "crack" you hear. Read on.

Sound doesn't always travel at the same speed (see p. 28). It depends, for one thing, on what medium it is traveling through. The pressure waves can't be transmitted from one place to another unless the transmitting substance has molecules that can collide with one another effectively and pass the energy on.

Suppose we have two trains on the same track, colliding head-on. (DO NOT TRY THIS AT HOME!) The impact energy will be transmitted, car by car, down the lengths of the trains, from their engines all the way to their cabooses (unless they derail, of course). Each car transmits its shock to the next car in line by colliding with it; that car transmits it to the next one in line by colliding with it, and so on, and the shock energy travels down the trains like a wave. That's how the pressure waves of sound are transmitted through materials, but by collisions of molecules, rather than railroad cars.

You can see that if the railroad cars weren't coupled very tightly together it would take more time for the shock wave to travel all the way to the cabooses, because time would be lost by each car's having to move toward the next car before it could collide with it. In the same way, it takes more time for a sound wave to be transmitted through a substance if the molecules of that substance aren't very close together.

In air, as in all gases, the molecules are very far apart, so sound travels relatively slowly through air: about 900 miles per hour (1,400 kilometers per hour) at sea level and room temperature. In water, the molecules are much closer together; sound travels through water at 3,300 miles per hour (5,300 kilometers per hour). In a dense solid such as steel, it travels at 13,000 miles per hour (21,000 kilometers per hour).

So much for how *fast* sound travels. Now let's look at how it changes as it travels.

As you can imagine, the close-up sound of lightning is a sharp, high-pitched crackle—just what you'd expect from a huge spark. But by the time a distant thunderclap reaches you, it may be a low-frequency rumble. The conclusion we draw from that is that low-frequency sounds travel longer distances than high-frequency sounds, which tend to peter out with distance. Ever notice that when your idiot neighbor plays his stereo loud enough to peel the paint off the walls you hear primarily the bass notes? The treble notes just don't carry as far and are also absorbed better by the walls. The reason is that the higher-frequency sounds are making the air and the walls vibrate more times per second, so they are using up their energy faster as they go.

That's why the low frequencies of the thunderclap carry farther than the high-pitched pops and crackles, and the farther away you are from the actual electrical event the lower the sound pitch will be. That's another way of comparing the nearness or farness (why isn't that a word?) of lightning strikes. The farther away the strike is, the later and lower will be the sound.

You must have noticed that thunder isn't simply high- or low-pitched, but is a mixture of high- and low-frequency sounds. That's because the lightning itself happens at a mixture of distances from you. The bolt may be miles long, with huge branches spreading out from the main stroke, so various parts of it are various distances from you, and that spreads out the frequencies of the sounds you hear.

You have also noticed that thunder rumbles and rolls for

an extended period of time. There are two reasons for that. One, the sound is traveling various distances from the various branches of the bolt, and two, it is echoing off the ground as it travels.

Now you may crawl back under the bed.

NITPICKER'S CORNER

Sound waves aren't transmitted through air simply by making the air molecules collide with one another in a straight line, like a string of railroad cars in a crash. Sound energy converts "smooth air" into a series of zones that are alternately compressed and expanded. That is, sound forces the air into alternating regions of high and low density. It is these density alternations that hit your eardrum at the rate of a certain number of compressions and expansions per second. The more of these compressions and expansions that hit your eardrum per second, the higher the frequency, or pitch, of the sound that you hear.

The speed of sound in air varies quite a bit depending on the air's temperature and pressure. The rule of thumb I gave above for timing how far away a lightning bolt struck is only a rough guide, because we can't know the temperature and pressure of the air where the bolt created most of its thunder noise or the air conditions between there and us. I chose four seconds for each mile of sound delay, but you'll see five seconds suggested in other books. Don't sweat it. As mentioned above, lightning bolts are long, and they may create thunder all along their paths in air that has a variety of temperatures and pressures and is at various distances from you. That's why you may have trouble timing the thunder anyway; do you time from the flash to the beginning of the rumble, or the end? It's far from an exact science, unless we know a lot more about the lightning bolt than we usually do.

Fool Moon

Why is the moon so much bigger when it's rising and setting, compared with its size when it's high in the sky?

Practically everybody has noticed this oddity at one time or another. When the moon is low, near the horizon, it looks huge compared with how it looks a few hours later when it is higher overhead. The effect is especially noticeable when it is a big, beautiful, full disk—a full moon. But you can notice the effect at any phase.

People have been wondering about this curiosity for at least two thousand years, since long before they even knew what the moon was or how it moves around Earth. (But *you* know, don't you? Any doubts? See p. 168.) Now would you believe that in today's so-called space age we can play hopscotch on the moon, but we still don't know the answer to the puzzle about its apparent size?

As you can imagine, people have come up with dozens of "explanations" over the years. But all save a few of them can easily be shown to be wrong.

A definitive explanation of the Moon Illusion—and that's what it is, an optical illusion—continues to evade science. If it were a matter of physical science, I assure you we'd know what's going on by now, because physics is a highly advanced science. But apparently it's a matter of human perception, and our understanding of our own psychology isn't nearly as advanced as our understanding of the world around us.

If there is one thing we *are* sure of, it's that as it orbits around Earth, the moon certainly does not yo-yo up and down in size like a fat lady on a fad diet. Earth's original satellite isn't one whit bigger when it's rising and setting near the horizon than it is when it's directly overhead. So it's got to be something about the way it appears to our human eyes and brains. But what?

Before we shoot down some of the wrong theories and add our support to some of the more plausible ones, let's prove to ourselves that it is indeed an illusion—that when we think we're seeing bigger and smaller moons, we're really not.

TRY IT

Check your daily newspaper for the date of the next full moon; it's right there with the weather map. Or call your local TV meteorologist. On that fated night, go out as soon as it's dark and sink your fangs into the creamy, white throat of a beautiful young . . . Oh, sorry. Wrong book.

On that fated night, go out as soon as it's dark and locate the moon while it's still low, near the horizon. If you have to, go to the nearest hilltop. Now take out the ruler that I forgot to tell you to bring along, hold it at arm's length against the moon and measure its apparent size. It will span about a half-inch (12 millimeters). Write down its "size" to the nearest sixteenth of an inch (or millimeter).

Now wait a few hours until the moon is high in the sky and measure its apparent size again. What did I tell you? It's exactly the same, isn't it?

or . . .

Take several pictures of the full moon when it's near the horizon and later, as it climbs higher in the sky. Use a telephoto lens to make a large image on the film. If you have a zoom lens, <u>make sure</u> you're shooting all pix at exactly the same focal length. Use several shutter speeds to get at least one good exposure at each position. You'll find that the moon is exactly the same size in all the pictures!

So rulers and cameras aren't fooled, but we Homo sapiens are. Humbling, isn't it?

Now to shoot down some of the theories that have been advanced.

"When the moon is low, you're unconsciously comparing it with trees, buildings and mountains on the ground, and it looks big compared with them. But when it's all alone up in the sky there's nothing to compare it with, so you don't think it's so big."

Well, maybe. But even out on the prairie, where there's nothing at all on the horizon, it still looks bigger when it's low.

"When the moon is low, you're seeing it through a lot more air than when it's directly above. All that air can act like a lens, refracting (bending) the light rays like a magnifying glass."

Sorry, Charlie, but any such refraction effect is small and can make the moon look slightly distorted in shape, but not in size.

"When the moon is low you're looking straight ahead, but when it's high in the sky your head is tilted upward and your eyeballs are slightly squashed, and that makes . . . yada, yada, yada."

Nonsense.

So what's the answer? Psychologists who study human perception have a couple of fairly convincing theories.

Theory number one: All our experience since the day we opened our eyes (some psychologists think it's even inborn) has taught us that when an object is coming toward us it gets bigger. Think of an approaching airplane, or even a fly ball coming toward you in the outfield. But the "moon ball" seems to be breaking all the rules; as it moves overhead it isn't getting any closer and it isn't getting any bigger. So your brain interprets it as being unnaturally small, and that's the conclusion you draw. It's not that the horizon moon looks bigger, it's that the overhead moon looks smaller.

I'd be more inclined to believe that theory if the moon rose in a matter of seconds. When I glance at it high in the sky, I'm really not comparing it with what it looked like several hours earlier.

Theory number two: Look up at the sky. If you didn't know better, wouldn't you think it was a huge, overhead dome? Ancient astronomers, in fact, thought that it literally *was* a dome, into which the stars and planets were set like jewels. Even in this space age, we still seem to have a built-in impression of the sky as a dome. We can't grasp the idea of infinity, so we tacitly imagine that it has finite limits.

Picture it consciously for a moment. Now what if I asked you how far away the sky-dome is? You're very likely to feel that the edge of the dome that touches the horizon is farther away than a point on the dome that's straight overhead. In other words, we think of the sky as a somewhat *shallow* dome; it just seems more comfortable that way. Why? Our experience has always told us that horizons are far away, but there is nothing in our experience, and no visual cues or clues, to tell us that the "top of the sky" is also far away.

Thus, when the moon is near the horizon, we subconsciously believe that it is farther away than when it is over-head. *But all of our visual experience tells us that farther-away things look smaller.* So when the Man in the Moon thumbs his nose at our expectations by remaining his usual size even when he's "far away" on the horizon, our brain says, "Wow! That guy must be really big." And that's the impression we get.

My money rides on this last explanation.

NITPICKER'S CORNER

Everything I've said about the moon (except that it orbits Earth) goes for the sun too. It also looks bigger when it is near the horizon, and for the same reasons. Haven't you no-ticed those spectacular sunsets with those absolutely huge suns? Now you know why your pictures of sunsets are always so disappointing. ("I could have sworn it was much bigger than that!")

Twinkle, Twinkle Little . . . Planet?

Why do the stars twinkle?

The answer that you see everywhere is that the twinkling is caused by turbulence in the atmosphere, which distorts the light coming from the star. But that doesn't explain *why* "at-mospheric turbulence," whatever that is, should distort light in the first place, or where the on-and-off blinking effect comes from, or why only stars, but not planets, twinkle. (That's right. If that dot of light in the sky isn't twinkling, it's a planet or an airplane. The only star that doesn't twinkle is the sun. Why? Read on.)

Mere turbulence in the air, more commonly known as wind, has no effect whatsoever on light waves. The light is traveling at 671 million miles per hour (more than a billion kilometers per hour) (see p. 90), and it couldn't care less if the air it's passing through is poking along even at hurricane speeds of 100 miles per hour (160 kilometers per hour). What *does* distort light waves is the varying *temperatures* of the air, not its varying speeds.

Obviously, the temperature of Earth's atmosphere isn't the same everywhere. Not only are there varying climates, but the air's temperature varies a great deal with altitude (see p. 112). And that's not even considering the crazy quilt of hot-air patterns from sun-heated land, from factories and from politicians' promises that the starlight must penetrate before it can reach our eyes down here on the ground. The light from a star has to traverse a veritable obstacle course of air at different temperatures. Turbulence is involved only insofar as the winds are constantly scrambling the patterns of different-temperature air.

So what? Well, when light enters a transparent medium such as air, water or glass, it generally changes direction. (Techspeak: It is *refracted;* see p. 61.) That's how come those chunks of glass or plastic in front of your eyes can correct the way in which light is focused on your retina. But the *amount* that any given transparent medium will bend light depends on its atomic constitution. Air, for example, refracts or bends light less than glass does. But here's the punch line for all you twinkle fans: Warm air bends light to a lesser degree than cool air does. Although the atoms in warm and cool air are the same, they are farther apart in the lighter, thinner, warm air, so they can't do the refracting job as well. It's very similar to how warm and cold air bend sound waves (see p. 28).

Now any star (except the sun) is so far away that we see it as only a single, perfect dot in the sky, a geometric point with no apparent size at all, even when viewed through the most powerful telescopes. It looks as if it is sending us only a single ray of light. As that ray comes down to us through the atmosphere, it is scattered hither and yon as it passes through air of many different temperatures and bending powers. Whenever it is scattered away from our eyes, the star seems to disappear for an instant. That is, it blinks off. When the ray happens to scatter again into our eyes, it blinks on again. This on-and-off flickering is what romantics like to call a twinkle.

For a big-appearing object like the sun or the moon, all that light scattering doesn't matter, because there are so many light rays coming toward us that just as many of them are being scattered *into* our eyes as are being scattered *away* from them, and the image appears steady.

Planets may look as if they're absolute points of light like the stars, but they're not. Even a pair of binoculars will show them to you as disks. So they don't twinkle for the same reason that the sun and moon don't: While some of their light rays are being scattered away from our eyes, there are enough others coming toward us to keep the image steady.

And besides, "Twinkle, Twinkle, Little Planet" doesn't have the right rhythm.

You Didn't Ask, but . . .

Why do distant objects seem to ripple and shimmy on a hot day?

For much the same reason that stars twinkle, except that there are enough light rays coming from the object that no matter how much they scatter, some of them will always be reaching your eyes. So there's no actual twinkling.

When you look down the road on a hot day, you may see shimmering "lines of heat" or "heat waves," and a distant car will appear wavy. What you're seeing is the effects of light refraction: the bending of light rays when they leave one transparent medium and enter another.

In this case, the light rays from the car you're looking at

are passing through various regions of air on their way to your eyes—air of different temperatures and different light-bending abilities, depending on just how hot each section of the road happens to be. A light ray coming at you from one part of the car may be traversing a different combination of air temperatures—and hence may be bent by a different amount—than the light from some other part of the car. And that looks to you as if the car itself is bent.

But why does the distorted image keep wavering? Because the rising hot air and other air circulations keep changing the patterns of air temperatures through which the light is traveling. If the consequent amount of ray-bending keeps changing, so does your image of the car.

Man in Moon Moons Earth!

How does the moon manage always to keep its same face toward Earth?

Sounds odd, doesn't it? Either it's the most colossal coincidence that ever occurred, or there's something real fishy going on. Well, even the fishiest-seeming coincidences can have rational explanations.

Your first guess might be that the moon isn't spinning on its axis the way Earth is, and that it just goes around us, maintaining the same orientation toward us. But it is spinning. And even if it weren't, we would still be seeing all sides of it as it circled Earth. Here's why.

TRY IT   Let's say that you're the moon and your buddy is Earth. Stand several feet away, facing him. Now keep staring at the same spot on the wall—that is, don't spin on your axis—and circle around him. (In square dance parlance, perform a do-si-do.) Notice that at some time during your circling, you can't

help showing him your backside. To avoid that,
you'd have to keep facing him all the way around,
and that requires that you rotate one full turn.

Then if the moon is indeed rotating while circling Earth,
and yet keeps the same side always facing us, it must be turning at a perfectly synchronized rate: exactly one moon turn
for each circle that it makes around Earth. How in the world
can that happen?

Well, you know that the moon and Earth are tugging on
each other gravitationally. You also know that the moon's tug
on Earth pulls the oceans up into bulges called tides (see p.
165). But what you probably never thought of is that Earth is
also pulling up bulges on the moon—not bulges in its nonex-
istent oceans, but bulges in the moon's very ground. Slight
bulges, to be sure, but bulges nevertheless. Call them tides in
the ground, if you wish.

Remembering that Earth's pull on the moon is much stronger than the moon's pull on Earth because Earth is so much more massive (see p. 104), you will realize that Earth's pull can deform the moon a lot more than the moon's pull can deform Earth.

This deformation of the moon by Earth's gravity acts like a brake on the moon's rotation. It's as if Earth's gravity were trying to hold on more tightly to the moon bumps because they're a tiny bit closer. And that has a slowing-down effect. So even if the moon was spinning like a top billions of years ago, Earth's gravity has slowed it down to its present crawl. We have grabbed the moon, tamed it and made it pirouette to our own tune.

And by the way, the moon is doing the same thing to our home planet, albeit to a much lesser extent because its gravitational pull is weaker. That is, by tugging on the oceans, the moon has been slowing down Earth's rotation, making our days longer. About 900 million years ago an Earth day was only eighteen hours long. Back then, the labor unions were joyful, because anything more than six hours of work was considered overtime. White-collar workers weren't so happy, though, because their annual salaries had to stretch over a 487-day year.

NITPICKER'S CORNER

Let's clean up a couple of points about the moon.

First of all, the moon doesn't show precisely the same face to us all the time. Although it keeps spinning at a constant rate, it wobbles a little bit from left to right, teasing us periodically with a glimpse of its backside. Mooning us, so to speak.

More surprising, perhaps, is the fact that if you really want to be picky about it, the moon doesn't orbit around the center of Earth. That is, the center of the moon's orbit does not lie at Earth's center. The reason for that is that gravitation is not a one-way street, with Earth holding the moon in orbit. The moon also holds Earth, but not as strongly, of course, because of its smaller mass. You might say, then (and I will), that Earth is trying to orbit the moon to a slight extent. The result is that they're each orbiting the other in a sort of whirligig dance.

It's like two square dancers, a heavy man and a light woman, executing a swing-your-partner maneuver. Each is orbiting around the other, but the woman, being lighter, does most of the orbiting. Somewhere in between them there's a fixed point that isn't going in circles at all; it is the nonmoving center of both orbits. That point will be closer to the man than to the woman, because being heavier, he is a better anchor for the whole whirling configuration.

We call that stationary point the *center of mass* (see p. 104) of the couple.

It's the same with the dance of Marilyn Moon with Ernie Earth. The center of both orbits—the center of mass of the Earth-moon system—will be much closer to Mr. Earth than to Ms. Moon. In fact, Earth is so much heavier than the moon that the center of mass will actually lie somewhere within Earth—somewhere outward from its geometric center.

To sum up: Instead of saying that the moon orbits Earth, we should really say that the Earth-moon system revolves around its center of mass.

It's the Moon, Stupid

What makes the ocean tides? I know, it's the moon. But how? And why are there two high tides and two low tides every day, when there is only one moon?

Whenever someone pompously proclaims that the ocean's tides are caused by the moon, everybody mutters, "Uh, okay," and goes away just as puzzled as before.

"It's the moon" is a cop-out, because a real explanation requires a lot more than that. Ocean tides are the net result of several forces produced by motions of the moon, the sun and Earth itself, all interacting in a complex way, but all very thoroughly understood by oceanographers and geologists.

Come with me, and we'll sort it all out. Well, most of it, anyway.

Picture Earth and the moon as two balls, with the smaller moon-ball circling the Earth-ball more or less around the equator. But stop the motions of Earth and moon for a moment while the moon is to the right of Earth. Got the picture? Earth left, moon right.

The moon's gravitational force is trying to pull the center of Earth toward its own center—toward the right. (Why the centers? See p. 104). Let's call this attraction the "center-to-center pull." But on Earth's right-side oceans, the pull is slightly stronger than the center-to-center pull, because the right-side oceans are closer to the moon than Earth's center, and gravitational force is stronger at closer distances (see p. 23). This slightly stronger pull raises the oceans with respect to the rest of the planet, making them bulge outward, and we have a high tide on the side of Earth facing the moon.

Meanwhile, on the side of Earth opposite the moon (the left side), the oceans are slightly *farther* from the moon than Earth's center, and they therefore feel a slightly *smaller* rightward pull than the center-to-center pull. The stronger center-to-center pull pulls Earth slightly away from the left-side oceans, and the oceans are left bulging out with respect to the rest of the planet. That makes a second high tide on the opposite side of the world from the moon.

Thus, there are always two bulges of ocean water on opposite sides of Earth—on the side facing wherever the moon happens to be at the moment, and on the directly opposite side.

Now let's permit Earth to rotate. As it spins merrily beneath the two bulge-making forces, each location on Earth passes through a high-tide situation twice in its twenty-four-hour rotation, giving each location two high tides per day. And in between the high tides, what else? Two low tides. After all, the high-tide water has to come from someplace.

And by the way, if you're not too selective about whom you listen to, you may have heard someone say something like this: "Humans are more than half water, and since the moon acts on water to make tides, the phases of the moon affect human behavior."

Well now, look here. The oceans of the world weigh 1.5 billion billion tons and are moved only a few yards (meters) by the moon's gravity. A human body might contain a few hundredths of a single ton of water. Gravitational forces are proportional to mass; figure it out. Anyone who believes that the moon's gravity can affect human behavior must have water on the brain.

NITPICKER'S CORNER

The two daily high tides aren't exactly twelve hours apart. At any given location on Earth, they're twelve hours and fifty minutes apart.

Why? Because the bulges are caused by the moon's attraction, and they move along with the moon in its travels around Earth. While Earth makes its full eastward rotation (see p. 119) in a period of twenty-four hours, the moon is also moving eastward, so it gets slightly ahead of any given location on Earth. Earth then has to rotate an extra fifty minutes in order for that location to catch up with the moon—that is, to catch up with the next high-tide bulge.

Another important nit to pick: The tides aren't caused only by the moon. There's another big thing out there with an awful lot of gravity—the sun. It's 27 million times more massive than the moon, but it's 397 times farther away. The way gravitation works, distance reduces the force a lot more powerfully than mass increases it. (Techspeak: *The force of gravity increases in direct proportion to the mass, but it decreases in proportion to the square of the distance.*)

It works out that the sun's gravitation affects the tides about 46 percent as strongly as the moon's does. Tracing the subtle effects of that 46 percent on the tides would be a lot more work than either you or I care to do. Ignoring the sun's effects still gives us a pretty good understanding of the tides.

Time and Tide Wait for New Moon

Why are the high tides higher when the moon is full?

It's easy to fool yourself into thinking that the moon is bigger when it's full, and that it therefore pulls on the oceans more strongly to make higher tides. But the moon is always the same size and distance away as it circles Earth. It is just lit up differently by the sun at different times in its journey. That's why it looks to us like a whole disk (a full moon), a partial disk (a semicircle or a crescent) or no disk at all (a new moon). In other words, it goes through phases.

When the moon, sun and Earth happen to be all lined up, we see either a full moon or a new moon. The moon looks full when Earth is in the middle, between the moon and the sun. Think of it as if we are sitting in Theater Earth, with the Man in the Moon on the stage and Spotlight Sun behind us. We'll see the full face of the Man in the Moon. On the other hand, when the moon circles around behind us Earthlings, getting between us and the Spotlight Sun (turn around in your theater seat and look at the moon behind you) we see the moon as a darkened disk—that is, a new moon.

In either of these lined-up arrangements—full moon or new moon—the sun's and moon's gravitational forces are pulling along the same line of direction, and they reinforce each other to produce an extra-high tide: a "spring tide." The name has nothing to do with the spring season; it's called that because it "springs up" twice in every moon cycle: about every two weeks.

Blue Moon, Part One

How often is "once in a blue moon"? Does it have anything to do with the real moon?

There are two answers to the latter question: No and Yes.

The No answer: The expression "when the moon turns blue" was used for hundreds of years to mean "when hell freezes over" or "fat chance." "Blue moon" first appeared in print in the nineteenth century, but was probably used even before that because it's a quirky idea and almost rhymes. There was no intention to connect either of these expressions with the moon's actual behavior. (But people might once in a while have seen a real, blue-tinged moon caused by smoke in the air; see p. 170.)

The Yes answer: Whenever there are two full moons in the same month, the second one is often referred to as a blue moon. Calling it that is a very recent development. It dates from a March 1946 article in the astronomy magazine *Sky and Telescope,* based on an article in the *Maine Farmer's Almanac* that had appeared ten years earlier. The editors of *Sky and Telescope* have recently admitted, however, that they misinterpreted the *Maine Farmer's Almanac* article and that the title "blue moon" was really meant to be bestowed upon the fourth full moon in any season. Seasons are three months long, so they would ordinarily have only three full moons.

That makes a big difference. The fourth full moon in a season is not necessarily the same full moon as the second one in a month; it might happen to fall in a month all by itself. But the fourth-one-in-a-season concept is not as easy for people to grasp as simply counting the number of full moons in a month (anybody can count up to two), so I predict that the second-full-moon-in-a-month "blue moon" will never die, no matter what the astronomers say.

It isn't very unusual for two full moons to fall in the same month; it happens about four times a year, much more frequently than a fourth full moon in a season, which really does happen only once in a blue moon—every two and a half years or so.

Here's how two full moons can occur in a single month.

As you know, our calendar contains eleven 30- or 31-day months plus February. But the lunar month, the time it takes the moon to circle Earth (you know, of course, that it does; see p. 162) and return to the position in which it is totally illuminated—full-looking—is only about 29½ days. So two of those 29½-day illuminations can easily fall within the same 30- or 31-day period. It can never happen in February, though, because at only 28 or 29 days, February is shorter than the lunar month.

Blue Moon, Part Two

C'mon, now. Does the moon ever really turn blue?

Yes, but only once in a . . . great while. There has to be exactly the right kind of smoke or dust in the air.

It happened most spectacularly in 1883, when the Indonesian volcano Krakatau blew its top, spewing dust all around the globe. The bluest moon since Krakatau was caused by a series of forest fires in western Canada in 1951. When these things happen, the moon itself doesn't change color, of course; it's just the way it appears when viewed through the smoky air.

Understanding this will take us a bit away from astronomy, but the explanation involves some fundamental ideas about the nature of light that will serve us well in many other situations. So even if you don't care a fig about sad-looking moons, stick around.

What's behind a blue-appearing moon—and a lot of other things that we see—is the fact that light scatters. I don't mean that it reflects, such as when it bounces back off the bathroom mirror to remind you that you're getting older. By "scattering," scientists mean that individual particles of light bounce off individual molecules and other tiny particles, like billiard balls bouncing off one another.

Did I say *particles* of light? Yes, indeed. And you thought that light consisted of waves? *Waves* of energy, rather than *particles* of energy? Well, we're both right. Let's get that little problem out of the way first.

Light—and all other so-called electromagnetic radiations, from radio waves to X rays—are indeed waves of pure energy, traveling through space at the speed of, uh, light. We can manipulate light waves by putting them through specially shaped pieces of glass or other transparent materials: lenses and prisms. Practitioners of the science of optics, who bring us our microscopes, telescopes and eyeglasses, have no problem treating light rays as if they were pure waves, making them reflect and refract (change direction) to perform a variety of useful optical tricks.

But certain other things that light does, such as knocking electrons out of atoms, can only be explained if light consists of a stream of tiny particles, like bullets from a machine gun. We call those light bullets—and the bullets of other electromagnetic radiations—*photons.*

So is a beam of light a stream of waves or a stream of particles? Perhaps the most astounding and unsettling discovery in human history was that light and other electromagnetic radiations behave as if they are *both* waves and particles. Or, if you prefer, they behave as *either* waves or particles, depending on what you catch them doing at any particular moment.

When a man named Albert Einstein (1879–1955) proposed in 1905 that light can knock electrons out of atoms as if it were a stream of bulletlike particles, he earned a Nobel Prize. (His prize was awarded for this work, for explaining this so-called photoelectric effect, not for his theories of relativity, which came much later.) It was almost as hard for physicists to swallow this two-faced idea as it is for you. But plentiful evidence has since proven beyond any doubt that it's true. Not only that, but (are you ready?) the reverse is also true: Honest-to-God particles such as electrons can act as if they are waves.

Physicists are now quite comfortable with this weird subatomic schizophrenia and refer to it as *wave-particle duality,* or simply *duality.* No amount of further palaver on my part will make it seem any more reasonable to you. That's just the way it is, and if you don't like it, move to another universe.

Didn't mean to be brusque there, but we've got to move on and explain blue moons.

I said that they're caused by the scattering of light photons, presumably after colliding with something. Well, what would cause a particle of light to veer off in a different direction after a collision? Obviously, a collision with some other particle that is at least as big as it is. Because certainly, a baseball wouldn't be scattered by a collision with a mosquito, would it? But if it collided with another baseball during its home-run flight out of the ballpark, it would be deflected into some much less fortuitous direction.

So we must conclude that a photon of light will be scattered best when it collides with something that is approximately its size.

But what is the size of a photon? How do you measure it, when it won't even stand still, oscillating like a wave whenever it feels like it? Well, if light can be schizophrenic, so can physicists, who take refuge in the *wave* description of light whenever *they* feel like it. They consider the "size" of a photon to be its *wavelength* when it's acting as a wave. (As a wave oscillates up and down, which is what waves do, the wavelength is the distance between two successive "ups" or two successive "downs.") Our conclusion, then: Light will be scattered best from objects that are approximately the same size as its wavelength.

Hold on, now. The moon is about to turn blue.

The light that comes to us from the sun is a mixture of all colors—all wavelengths from red, the longest, to violet, the shortest. When all the daylight colors are mixed together, as they are when we receive them here on Earth, our eyes and brains interpret the light as no color at all: white light (see p. 38). That's the light that we can see. But there are other "colors"—infrared and ultraviolet, for example—that our human eyes are simply insensitive to.

In the light that we can see, blue has just about the shortest wavelength; it consists of the "smallest" photons. It will therefore be scattered best by the smallest particles that it may encounter in its travels through the air, namely the molecules of nitrogen and oxygen that the air itself is made of. It was Einstein (again) who figured out exactly how molecules scatter light of different wavelengths: The shorter the wavelength, the more the scattering.

Now what if the air contains some bigger-than-molecule-sized particles, such as particles of dust or smoke? Then the other colors of light, the longer wavelengths, can be scattered more than usual. If—and it's a big if—a forest fire or volcano should happen to make smoke or dust particles that are exactly the right size to scatter longer-wavelength red light, then the light coming down from the moon will have a lot of its red scattered away before it reaches the ground. And light that is deficient in red looks bluish to us. Hence, so does the moon.

You Didn't Ask, but . . .

Why is there always a blue haze around some mountains?

Evergreen trees give off vapors of resinous chemicals. These vapors can react with ozone in the air to produce extremely tiny solid particles of just about the right size to scatter blue light. So blue light photons are being scattered and rescattered all around, while the other colors are passing by in straight lines. Thus, more blue reaches our eyes than the other colors.

You Didn't Ask This Either, but . . .

Is that why the sky is blue?

Pretty much, yes. But the sky isn't blue because the blue light is being scattered by dust, as was once believed, and as many people still believe. The blue light is being scattered by the nitrogen, oxygen and other molecules that make up the air. These molecules are best at scattering the shortest wavelengths, with blue light being scattered about ten times more than red. When you look up at the sky, we're seeing all that extra blue light that may not have started out in your direction, but that has been scattered and rescattered into it.

NITPICKER'S CORNER

Photons of light will also bounce off much bigger things—things much bigger than their wavelengths. That soaring baseball a couple of pages back would certainly be deflected (and its batter dejected) if it encountered the outfield wall in its attempt to become a home run. So wavelength doesn't matter when the scattering object is bigger than *all* the wavelengths in visible light; they'll *all* bounce off. That's what happens when all the colors of light are reflected equally from a solid surface such as a mirror. No change occurs in the color balance.

Is It Cold Up Here, or Is It Me?

Why is it so cold in space?

It isn't. Satellites and space shuttles do indeed get cold up there, but it's not because it's cold up there.

First of all, there's really no such thing as cold, no matter what the penguins tell you. Cold is a linguistic concept, not a scientific one. Our caveperson ancestors needed a word for "not hot," and "cold" (or its grunt equivalent) is what they came up with. It's like light and dark, wet and dry. Light and water are tangible things, but dark and dry denote the *lack* of light and water. They're negative adjectives, if grammarians will permit me.

Okay, that was semantic fun, but everyone knows what we mean by "cold." So explain why space isn't cold, already.

Okay, okay.

Heat is energy. It's the energy that an object's molecules have by virtue of the fact that they're in motion (see p. 79). Why are they in motion? Because around 12 billion years ago an incomprehensible amount of energy arose in the void (or whatever) via the Big Bang—that mind-boggling blast that scientists believe ignited the universe—and all the atoms are still quivering. Some, the hotter ones, are quivering more than others; we refer to those others as colder.

Some forty years ago, when we left the cuddly atmosphere of our native planet to venture into the vast beyond, we encountered for the first time an environment in which there is no heat to compare anything with because there are no (or precious few) molecules up there to quiver, and the word "cold" became even more meaningless. Space can be neither hot nor cold in any sense of the words, because it is empty of matter.

Then why do satellites and spacecraft get so . . . frigid? Parts of NASA's space shuttle do get down to a couple of hundred degrees below zero Fahrenheit (around −130 degrees Celsius).

Here's what's happening. A space shuttle or any other object can gain or lose heat not only by being in contact with stuff that's hotter or colder—and that's out because there's no stuff up there—but also by *radiation*. The sun and stars are putting out all sorts of radiation—waves of pure energy, both visible to the human eye (light) and invisible (ultraviolet, infrared and others). This radiation travels through space without being diminished because there's nothing there to absorb it. But when it strikes an object, for example a space shuttle, some of it will bounce off and continue on its way in a different direction. Some of it will be absorbed, however (see p. 37), and its energy will dissipate into heat. Thus, the space shuttle is receiving radiated heat from the sun and stars. The sun, of course, is by far the chief heat radiator because it is so much closer than the other stars.

But at the same time the shuttle, still carrying its burden of earthly warmth, is radiating some of its own energy away, because anything that has any warmth at all sends out infrared radiation—"heat radiation" (see p. 112). That's how night-vision devices can "see" people in the dark: by the infrared radiation they're sending out. And that's how old-fashioned radiators work: They radiate heat into the room, rather than blowing hot air around the house.

The shuttle, then, is receiving lots of radiated heat on the side facing the sun while radiating heat rapidly away on the other side, which then gets exceedingly cold.

Note, then, that the shuttle itself can be said to be cold because it is a real object, but the environment it is flying through is not cold, either semantically or physically.

BAR BET   It's not cold in outer space.

6

All Wet

It's the one substance that is indispensable to all living things.

It makes up more than half of our own body weights.

It is the most abundant chemical on Earth, with more than a billion billion tons of it covering 71 percent of the planet's surface and probably another billion tons in those little plastic bottles that everyone carries around these days.

When even a little bit of it is discovered on another planet, astronomers grow giddy with speculation about the existence of extraterrestrial life.

It is water, H_2O, one of the simplest and most stable of all chemical compounds.

We normally think of water as a liquid, because that's what it is throughout our most comfortable range of living temperatures: between, say, 40 and 80 degrees Fahrenheit (4 and 27 degrees Celsius). But as you know, at any temperature below 32 degrees Fahrenheit (0 degrees Celsius) it prefers to exist as the solid that we call ice. And at any temperature above 212 degrees Fahrenheit (100 degrees Celsius) it prefers to exist in the form of vapor—an invisible gas, just like the nitrogen and oxygen gases in the air.

(Water vapor isn't steam. Steam is a cloud of tiny droplets of liquid water that are too small to fall out of the air.)

Water doesn't have to reach its boiling temperature in order to turn at least partially into vapor. Wherever there is water there is water vapor in the air around it. We sometimes call it humidity, and it has far-reaching consequences in many aspects of our lives, far beyond making us uncomfortable in the summertime.

In this chapter we'll look at some of the amusing things that water does when in its liquid form, such as making coffee stains, making the oceans both salty and blue and making your wet shower curtain slap you right in the . . . shower. We'll take a small detour through the Panama Canal on our way to the kitchen, where we'll play with some ice cubes and lollipops before going to the laundry to find out what's inside all those detergent bottles and boxes. Then we'll examine how water vapor affects cosmetics, clothes dryers and that old devil humidity.

And as usual, we'll explode a few misconceptions along the way, this time concerning the color of water, the flow of glass and whether warm air can really "hold moisture."

The Watery Blues

Why is the ocean blue? Is it just a reflection of the sky?

No. That's a common belief that just doesn't hold water, so to speak. First of all, the ocean's surface isn't exactly what you'd call a mirror. And second, how come it's a much darker blue than the sky?

No, the world's oceans really and truly are blue—many different shades of blue (ask any sailor), depending on several factors, a few of which we'll talk about below.

But here's a surprise: Even crystal pure water—without the salt, silt and fish—is blue. That's in spite of the fact that almost every dictionary defines water as "a colorless, odorless liquid." All you have to do is fill your bathtub and see for yourself that it isn't colorless.

TRY IT Fill your bathtub and look at the water. You'll see that it's a pale blue color. (I assume that your bathtub is white.) The only reason that you don't see blue in a glass of water is that you're not looking at <u>enough</u> water. The color builds up, or accumulates, as you look into or through thicker and thicker layers of it. If the windowpanes in your house were ten times as thick as they are, you'd see that the "colorless" glass is really green.

So why is water blue? Because when daylight, which contains all colors of light mixed together (see p. 38) hits the water and penetrates it, certain colors in the daylight are absorbed by the water molecules (see p. 38). The light that is reflected back from the bathtub and reaches your eye after passing through the water is then diminished in those particular colors, so it has a different color composition from the original daylight.

Specifically, water molecules show a slight preference for absorbing the orange and red portions of the sun's light. Light that is diminished in orange and red looks to us as if it has too much blue, compared with what we call "white light." So the water appears blue.

But an ocean is a much more complicated kettle of fish

than just H_2O. In addition to the obvious salts and minerals, it contains *plankton,* for one thing: tiny plants (*phytoplankton*) and animals (*zooplankton*) that are too small to settle out and that float perpetually about until decomposed by bacteria or eaten by anything bigger than they are. (It's a cruel world.)

Seawater also contains a lot of miscellaneous dissolved organic matter that scientists call by its German name, *gelbstoff.* Loosely translated, it means "yellow crap," because that's what it looks like when it's dry.

When daylight enters seawater, the phytoplankton absorbs mostly blue light plus a little red, while the *gelbstoff* absorbs mostly blue light. These absorptions shift the balance of the remaining light from the pale blue of pure water to a deeper, purplish blue. That's why the oceans are darker than the water in your bathtub, which I certainly hope is devoid of *gelbstoff.*

Unfortunately, the many faces that the seas present to us in different weather and in different parts of the world are not that easy to explain. For one thing, it's not just the *absorption* of light that gives seawater its color, it's also the *scattering* of light (see p. 170). Certain colors of light are scattered by microscopic particles of matter in the water. When a photon of light (see p. 170) hits one of those particles, which might be anything from a single molecule on up, it can ricochet off in another direction. This changes the distribution of colors that reaches our eyes.

It is just such a scattering of light from air molecules that makes the sky blue (see p. 174), because air molecules scatter blue light more than other colors. Some scientists have tried to explain the oceans' blueness as being due entirely to the same kind of scattering, but they've apparently never peered into their bathtubs.

Phytoplankton is an especially good scatterer of green and yellow light, so in general the more phytoplankton there is, the more greenish the water will be. That's what is largely responsible for the beautiful greenish turquoise color of the waters surrounding the Caribbean and South Pacific islands. The tropical climate and abundant sunlight there create lush breeding grounds for plankton.

And honeymooners.

All That Salt and No Popcorn

Why are the oceans salty?

When you say "salty," you're undoubtedly thinking of sodium chloride, common table salt. But to a chemist, a *salt* is any member of a large class of chemicals, and there are dozens of them in the oceans.

To put the word "salt" in perspective, please indulge me in a one-paragraph chemistry lesson.

A "molecule" of salt (it's not really a molecule in the strict sense, but I won't tell anybody if you don't) consists of a positively charged part and a negatively charged part that, being oppositely charged, attract each other. The positive and negative parts are called *ions*. In the case of sodium chloride the positive ion is a charged atom of sodium and the negative ion is a charged atom of chlorine. But a salt's positive ion can be a charged atom of any metal, and there are some eighty-five known metals. Also, there are many negative ions besides chloride, so you can see that there is a very large number of possible salts.

End of chemistry lesson.

The main metal ions in seawater are sodium, magnesium, calcium and potassium, while the main negative ions are chloride, sulfate, bicarbonate and bromide. Your question, then, is how all this stuff got into the oceans in the first place. The short answer is that it was washed out of the land by rainwater, which then flowed as rivers to the seas.

Seawater is continually being recycled. Each year, the top meter (3 feet) or so of the oceans evaporates into the air, moves around in various weather systems and falls back onto the oceans and land as rain and snow. Of this precipitation, 76 percent falls on the oceans and 24 percent falls on the continents. The water that lands on the continents flows down in streams and rivers, eventually returning to the seas. In the process of washing down, these waters pick up anything that will dissolve, mostly the salts that exist in the soils, rocks and minerals.

Any chemist will tell you that sodium salts dissolve more readily in water than do salts of potassium, magnesium, calcium or most other metals. More than any others, then, it's the sodium salts that dissolve and wash down into the oceans. There are approximately equal amounts of sodium and potassium in the soils, rocks and minerals, yet there is twenty-eight times more sodium than potassium in seawater.

All of these dissolved salts make up 3.47 percent of seawater, by weight. Only six elements make up more than 99 percent of those salts: chlorine, sodium, sulfur (in the form of sulfates), magnesium, calcium and potassium, in decreasing order.

Another source of sea salts is volcanic eruptions, both on land and under the sea, which spew out enormous amounts of solids and gases. Among the prominent volcanic gases are chlorine and sulfur dioxide, which may account for the fact that chlorine is the most abundant element in seawater, making up 55 percent of the salts' weight, while sulfates are second only to chlorides as the negative-ion portions of the salts.

Putting all this together, sodium and chlorine make up 86 percent of the salts in the oceans. So if you want to say that the oceans are salty because of sodium chloride, nobody will give you much of an argument.

You Didn't Ask, but . . .

Why are the oceans salty, but not the streams, rivers and lakes?

Rainwater washes down from the land into the streams, rivers and lakes, carrying dissolved salts just as it does when it washes into the oceans. But the difference is that the oceans are much older than the other waters—4 or 5 billion years old, compared with mere millions. Over those billions of years the oceans have been recycling their water—evaporating water that rains out on the land and flows back, returning each time with a fresh load of salts. These cycles have continually increased the load of salt in the oceans.

Where the Devil Is Sea Level?

I understand that the Panama Canal has locks because the Atlantic and Pacific Oceans aren't at the same level. Then what do we mean when we talk about elevations above "sea level"? Which sea?

Whoa! That's not why the Panama Canal has locks. The locks are there for the purpose of lifting ships over the hump of land known as the Isthmus of Panama. To get over that hump, the ships have to be floated upward 85 feet (26 meters) above the entrance ocean and then lowered back down into the exit ocean on the other side. That goes for both directions.

Why didn't they just dig a flat ditch from one ocean to the other, a so-called sea-level canal? Mostly because it would have meant excavating a tremendous amount of dirt at tremendous expense. But also, there would have been torrents of water gushing through a sea-level canal. That's not because of any permanent difference in level of the two oceans, however; their average sea levels are just about the same. It's because of the tides.

At the Pacific end of the canal the tides can rise and fall as much as 28 feet (5.5 meters), whereas at the Atlantic end the tides vary by only about 2 feet (60 centimeters). So there would be periodic surges of water through the canal from the Pacific to the Atlantic—that is, from east to west.

Do you think I got that backward? Isn't the Pacific Ocean at the *western* end of the canal? Nope. Because of the way the Isthmus of Panama snakes around, the Pacific entrance to the canal is 27 miles (43 kilometers) east of the Atlantic entrance. Check a map.

BAR BET A ship passing through the Panama Canal from the Pacific Ocean to the Atlantic Ocean sails from east to west. (It's actually southeast to northwest, if you must quibble.)

Now back to what we mean by "sea level."

Obviously, because of the tides, we can talk only about the average, or mean, level of any ocean at any location. While the mean levels of the Atlantic and Pacific Oceans may be approximately the same at the Panama Canal, for example, that doesn't mean that all the oceans in the world have settled down to a common level. You might expect them to be that way, because if you look at a globe you'll see that they're all connected; Earth's oceans are one gigantic swimming pool with chunks of land scattered about. But even when you average out the tides, there are reasons why the oceans differ in their levels.

For one thing, because gravitational effects are bigger for bigger masses (see p. 104), the bigger oceans will be raised into higher tides by the moon's gravitational pull (see p. 165). (Only the biggest lakes have tides.) Weather patterns also affect the sea levels. When the air pressure over an ocean is low, the water will actually expand. Moreover, prevailing westerly winds can actually make the water pile up somewhat toward the east. And finally, differences in ocean depth can have a gravitational effect on the water's level, because the deeper an ocean is, the more tightly its waters are packed down by gravity, and the lower its surface level will be.

These are all small effects, but when working on such huge bodies of water, they can make significant differences in the sea level at different locations in the world. Since they are all connected, the waters do try, of course, to seek a common level, but they are just too slow to keep up with all of these changing conditions.

So what is mean sea level? It's a carefully compiled average, measured over a period of nineteen years at many tidal stages in many places around the world. Whenever you hear that something is so many feet or meters above "sea level," or that the atmospheric pressure at "sea level" is so many inches or millimeters of mercury, it's understood that they're talking about mean sea level: a worldwide, long-term average.

Why-ing Over Spilt Coffee

When spilled coffee dries on my kitchen counter, it forms a brown ring, with almost nothing inside. Why does all the coffee go to the edges to dry?

For years, people have observed this phenomenon without giving it a second—or even a first—thought. Hundreds of less-than-fastidious, coffee-sipping scientists have probably glanced at the ring, mumbled something about surface tension and told their lab assistants to clean it up.

But it wasn't until 1997 that six scientists at the University of Chicago pondered this earthshaking question and published their results in the prestigious international scientific journal *Nature* for the benefit of all mankind—or at least for those slobs among us who don't wipe up their spills before they dry.

Here's what they concluded after producing reams of mathematical calculations, undoubtedly supported by lots of caffeine.

When a coffee puddle finds itself on a flat, level surface, it tends to spread out in all directions. In any given direction, the liquid will stop spreading when it hits a barrier, any slight irregularity in the surface that it can't cross, such as a microscopic ditch. Depending on where the barriers happen to be, the puddle will take on a certain shape: longer in this direction, shorter in that, like an amoeba.

As evaporation takes place, the puddle will start to dry first where it's thinnest: at the edges. That would have the effect of making the puddle shrink, pulling its edges back, but it can't do that because they're stuck in the ditches. So as water evaporates from the edges, it has to be replenished from somewhere, and the only place it can come from is the interior of the puddle.

Thus, there's a movement of water from the interior of the puddle to the edges, where it evaporates. That flow of water carries along with it the microscopic brown particles that give coffee its color. The brown particles then find themselves stranded at the edges when the puddle finally runs out of water.

TRY IT   First, clean your kitchen counter; no grease films allowed. If your countertop is light-colored, spill about a quarter-teaspoon (a milliliter) of coffee—black, no sugar—on it and let it dry overnight. You'll see the brown ring. If your countertop is dark, the effect is much better if you use salt water. Dissolve about half a teaspoon (a few grams) of table salt in half a cup (250 milliliters) of water and make a few quarter-teaspoon (milliliter) puddles on the counter. When they're dry, you'll see white rings of salt. The salt crystals are coarser than the coffee particles, so the rings will be more irregular.

Shower Power

When I'm taking a shower, why does the curtain sneak up and slap me on the leg—or somewhere?

You're lucky you asked that question today, because today is bargain day. I'm going to give you four answers for the price of one. It's not because I'm feeling generous, but because I can't make up my own mind about which one to believe.

To my knowledge, the National Science Foundation has not yet funded a comprehensive university research project designed to solve this perplexing problem, so scientists have been left to debate their theories over coffee and beer. Here are four contending solutions to the great shower curtain mystery. Ya pays yer money and ya takes yer choice.

(1) *Hot air rising.* The story goes that the air inside the shower is heated by the water and, as everybody knows, hot air rises. (But if you don't know *why* hot air rises, see p. 107.) If the hot air in the shower is rising, then cold air has to rush in to replace it at the bottom, and in the process it blows the curtain inward. This is a nice, simple, appealing and wrong explanation. Just try it with cold water instead of hot, and you'll see the curtain move inward just as much. (You don't actually have to *be* in the cold shower; you can do the experiment while standing outside.)

(2) *Electrostatic charge.* When water streams out of a narrow opening such as the hole in a showerhead, it can pick up a static electric charge. It's not too different from scuffing your feet on a carpet, whereupon some electrons are scraped off your shoes onto the carpet and you develop a positive charge (see p. 97). Electrons can also be scraped off—or onto—the water by the showerhead, depending on what it's made of.

But if the water's molecules were to pick up, say, a negative charge on their way out of the showerhead by picking up some negative electrons, those extra electrons would repel some electrons from the surface of the shower curtain, because similar charges repel each other. That would leave the curtain's surface with a deficiency of negative charge, and its inherent positive charges would dominate. The negative water and the positive curtain would then attract each other, as is the wont of opposite charges, and the curtain would move toward the water.

This isn't quite as far-fetched as it may sound. (Wait'll you see some of the other explanations.) Induced electrostatic charge, which is what the phenomenon is called, does happen and is well-known. Have you ever opened a carton packed with those styrene foam packing peanuts, especially when some of them are broken into small fragments? Just try to keep those maddening motes from jumping around or sticking to your hands as you try to brush them away. It's because of induced electrostatic charges.

TRY IT   On a dry winter day, put a few small fragments of styrene foam packing peanuts on a table. (If you haven't received a package lately, you can use torn-up scraps of lightweight paper.) Now walk across a nearby carpet while scuffing your feet to acquire a body charge. Go quickly to the table and try to touch the plastic peanuts. Even before you touch them, they will jump up to meet your hand. Your body's static charge induced an opposite charge in the plastic and the resulting attraction of opposite charges was enough to make them leap up toward you.

Whether or not induced electrostatic attraction is strong enough to move a shower curtain, however, is up for grabs.

(3) *Bernoulli's Principle.* The water is carrying along some entrained air, making an air current near the inside surface of the curtain. According to Mr. Bernoulli (see p. 18), the faster a gas moves across a surface, the lower its pressure

against that surface. Since there is no speeding airstream on the outside of the curtain, the air pressure on the inside is lower and the curtain moves inward.

(4) *The Coanda Effect.* Fluids have a tendency to stick closely to a curved surface over which they are flowing. This phenomenon is known as the Coanda Effect in honor of Henri Coanda (1886–1972), a Romanian aeronautical engineer who first called attention to it.

> **TRY IT** Hold a drinking glass horizontally in the stream of water from a slowly running faucet, so that the stream falls onto one side of the glass. Notice that when the water gets to the bottom it doesn't fall straight down. It sticks to the glass and follows its curved surface beyond the bottom before falling off.

In the shower, if the curtain is already curved inward somewhat, perhaps from one of the other effects, the water flowing over its surface may pull it farther inward because of Coanda stickiness.

NITPICKER'S CORNER

Figuring out exactly why a flowing fluid sticks to a curved surface took Coanda and other aerodynamic engineers more than twenty years. Here's the ultimate explanation.

The molecules of a fluid exhibit some stickiness toward one another; what some molecules are doing affects their neighbors because they are sort of tied to one another. (Techspeak: Fluids have a certain *viscosity.*) If one layer of a flowing fluid's molecules should have some adherence to a surface over which it is flowing, the rest of the molecules will be dragged down to the surface along with them to some extent, and the fluid as a whole will tend to stick more than we might expect.

In the case of the water on the glass, the first layer of water molecules wants to stick because water wets glass. (It doesn't wet wax, for example.) The second layer wants to stick to the first layer, so it is also weakly attached to the glass. The third layer sticks to the glass through the first two layers and so on, with each successive layer sticking less strongly than the preceding layer. Many other layers are dragged along for as long as the stickiness exceeds the pull of gravity, and then the water finally falls off the glass, having gone farther around the curve than we would have expected.

The attraction that air molecules have for one another is a lot smaller than in the case of water (Techspeak: The *viscosity* of air is a lot less than that of water), so it will stick to the shower curtain's surface a lot less, but the effect is still there. Both the water and its entrained air probably contribute to the attraction of the shower curtain.

That is, if you believe that the Coanda Effect is the true cause of your flapping, slapping curtain. Me, I favor the electrostatic explanation.

Psychotic Psocks

Unless I use one of those fabric-softening dryer sheets, all my clothes come out of the dryer full of static electricity, sticking to one another. What do fabric softeners have to do with static electricity?

Not much, except that the stuff in the dryer sheet happens to be good at both jobs. You can obtain the static-elimination function all by itself as a liquid in a spray can, so you can de-static your clothing even while you're wearing it without your having to—DO NOT TRY THIS AT HOME!—climb into the dryer.

The main ingredient in both types of products is a *surfactant,* a chemical that is made of what might be called bisexual molecules; they are attracted to both oil and water. Most other chemicals show a strong preference for one or the other.

For example, common salt (sodium chloride) is made of electrically charged atoms (Techspeak: *ions*), and charged atoms like to mix into—dissolve in—water because water molecules have electric charges that attract them. But salt won't have anything to do with fats and oils because their molecules don't have any attractive charged parts. Just try to dissolve some salt in olive oil and see how far you get.

Surfactants, however, are peculiar in that one end of each molecule is a fatty material that is attracted by oils, while the other end is charged and is attracted by water. Soap and detergent molecules are surfactants; their oil-loving ends latch on to oily dirt and drag it into the water by means of their water-loving ends. Or looking at it the other way, their water-loving ends drag water into oily places that it wouldn't ordinarily invade, thereby making the water wetter.

Now let's impregnate a paper sheet with a soapy-feeling surfactant chemical and throw it into the dryer along with our wet clothes. As they tumble, the clothes rub against the sheet and become coated with surfactant. The rather hefty fatty ends of the surfactant molecules impart a slippery, waxy feel to the clothes, "softening" them.

Then when the clothes begin to dry, their friction against one another rubs off some electrons and static electricity begins to build up (see p. 97). The charges can't build up as long as the clothes are wet because water conducts electricity well enough to conduct the rubbed-off electrons back to where they came from. When the water is gone, the charged ends of the surfactant molecules take over, conducting the charges away and killing any "static cling" that might result.

Deprived of their static cling, socks find themselves unable to bond with their partners and may suffer a severe separation-anxiety syndrome. In fact, a sock may become so depressed and emotionally unraveled that it will slink away through the vent tube in search of psychiatric help. That's why you will sometimes find a sock missing when you put away your laundry. I know you have wondered about that.

NITPICKER'S CORNER

There are three kinds of surfactants whose names you will see as ingredients on the labels of dryer sheets, clothes softener liquids, antistatic sprays and synthetic detergents (see the following). They may be listed as *cationic* (CAT-eye-ON-ic), *anionic* (AN-eye-ON-ic) or *nonionic* (NON-eye-ON-ic). The charged ends of the molecules can be either positively charged (cationic) or negatively charged (anionic). The nonionic surfactant molecules aren't charged at all, so they may be good at clothes softening but are of no use for killing static cling.

A widely used cationic surfactant is dimethyl ditallow ammonium chloride, and a common nonionic surfactant is polyethylene glycol monostearate. Laundry detergents (see the following) commonly contain the anionic surfactant sodium alkylbenzenesulfonate.

As if you cared, right? But now you can have fun decoding the fine-print ingredient lists on all those product labels. Run right down to the laundry and check them out.

Washday Wonders

Every laundry detergent claims to be "New," "Improved," "Unique" and better than the others. Aren't they all just soap?

No, those detergents aren't soap, although soap is a detergent. The word "detergent" simply means a cleansing substance, from the Latin *detergere,* to wipe off.

After more than two thousand years of using soap, which is easy to make by boiling up wood ashes with animal fat (don't you wonder how *that* discovery was made?), humans finally created *synthetic* detergents, which in many cases work even better than soap. Today we reserve the word "detergent" exclusively for those artificial chemical concoctions that take up so many acres of shelf space in our supermarkets.

All detergents, including soap, are *surfactants,* chemical compounds that have the knack of bringing oil and water together (see p. 192). Most dirt adheres to our skins, clothing, dishes and cars by means of a sticky, oily film. Coax the oily film into the water and you have succeeded in removing the "glue" that stuck the dirt to the objects.

But all those colorful bottles and boxes on the store shelves may contain a mad scientist's laboratoryful of other chemicals besides surfactants. Otherwise, how could the manufacturers keep claiming that their products are any different from or better than all the others?

Here is a list of what may be hiding in your laundry products, household cleaners, soaps, window cleaners, dishwashing detergents and the like, in addition to surfactants. And don't forget the most expensive ingredient of all: advertising. Lots and lots of advertising.

Acids and alkalis: Acids help to remove mineral buildup, while alkalis attack fatty and oily soils. Examples: acetic acid, citric acid, ammonia.

Antimicrobial agents: Kill disease microorganisms. Examples: pine oil, tricloban, triclosan.

Antiredeposition agents: Once you get the dirt off, you have to keep it from going right back to where it came from. Examples: carboxymethyl cellulose, polyethylene glycol, sodium silicate.

Bleaches: Remove stains and "whiten and brighten" your clothes. Examples: sodium hypochlorite (chlorine bleach), sodium perborate ("color-safe" bleach).

Builders: Counteract hard water, which interferes with the surfactant's performance. Examples: sodium carbonate (washing soda), sodium tripolyphosphate. The latter is one of the notorious phosphates in detergents. If phosphates get into the sewers and then into streams and lakes, they can harm the environment by disrupting the ecological balance.

The phosphates make algae grow in profusion, and when the waters can't sustain any more algae they die off, which provides a feast for bacteria, which use up oxygen in the water and kill the fish, which makes even more dead bodies for the bacteria to feed on, etc. Because of this, phosphates have been largely eliminated from detergents.

Corrosion inhibitors: Protect the metal parts of your washing machine or kitchen utensils. Example: sodium silicate.

Enzymes: Enzymes are natural chemicals that speed up natural chemical reactions. In laundry products they speed up the destruction of specific kinds of stains, such as grass. Examples: protease, cellulase.

Fabric softening agents: Soften fabrics and control static electricity. Example: quaternary ammonium compounds.

Fragrances: Cover up the smells of all the other ingredients and make you think your laundry is "fresh," whatever that means.

Optical brighteners: Make clothes look brighter by converting yellow light or invisible ultraviolet light into bluish or whitish light (see p. 34). Example: stilbene disulfonates.

Preservatives: Protect the product from oxidation, discoloration and bacterial attack. Examples: butylated hydroxytoluene, EDTA.

Solvents: Keep all the ingredients dissolved in liquid products. Examples: ethyl alcohol, propylene glycol.

Suds control agents: Control the amount of suds or make the suds keep their "heads." Examples: alkanolamides and—guess what?—soap.

Life in the laundry isn't as simple as when all we had to do was boil up a nice kettle of goat fat and ashes.

Glass Dismissed

My son's teacher told the class that glass is really a very thick liquid, and that given enough time you could see it flow under the influence of gravity. Really?

That's a commonly quoted "amazing fact" that is simply not true.

Liquids do get somewhat thicker as they are cooled, and because all glass was once a liquid while it was hot and being formed into shape, some people like to think that it gets thicker and thicker as it "supercooled," until it gets so thick that it acts like a solid. Well, the truth is that it *is* a solid.

If glass flows, its motion apparently requires more than four thousand years to become detectable, because that's how long glass has been around and nobody has yet come up with any convincing evidence of its motion. That's one hole in the "supercooled liquid" theory. But the bigger, absolutely gaping hole is that glass is *not* a supercooled liquid, despite the fact that quite a few textbooks and encyclopedias say it is.

The "supercooled liquid" fable has been around at least since I was in school and accepted everything my teachers said as gospel. But science and I have marched many a mile since then, and there is no longer any excuse for perpetuating the myth.

If you've ever observed glassblowing or the process of molding glass into shapes, you know that when it's hot enough the glass certainly does flow like a very thick (viscous) liquid. But as it cools down, we observe no sudden transition from liquid to solid, as we do, for example, when water cools down and turns into solid ice. That has led many well-meaning scientists to conclude that the glass must still be a liquid even at room temperature, because it didn't turn abruptly rigid. Moreover, the argument goes, solids are usually *crystalline*, meaning that their atoms or molecules occupy precise geometric positions with respect to one another, and glass's molecules don't. Examples of crystalline solids are ice, table sugar, salt or almost any mineral you can think of. If a typical solid's atoms and molecules weren't fixed in place, they could slip and slide over and around one another; in other words, they would flow like a liquid.

But a substance's molecules don't have to be in the form of a highly organized crystal in order for it to be a solid. There are such things as *amorphous* solids (from the Greek, meaning "without form"), in which the molecules are indeed fixed in place, but in a more or less random arrangement. That's the case with glass. It's a solid, all right; it's just not a crystalline one. (Techspeak: Its structure does not exhibit *long-range order.*) When glass is cooled from the molten state, its molecules can't find a repeatable, orderly arrangement into which to fit themselves. It's the same with many other amorphous solids such as plastics and lollipops. Translucent lollipops are sugar (sucrose) in an amorphous, glassy form, as distinguished from its crystalline form in the sugar bowl.

TRY IT Melt some sugar over very low heat in a small pan. If some sugar crystals stay on top without melting, stir them in with a fork. When it is all melted, but before it gets too dark, pour it out onto a cool, flat surface such as a tile countertop or the bottom of a metal frying pan. The sugar molecules will cool so fast that they don't have time to arrange themselves into crystals and will wind up in a glassy form. After it cools, you can eat your caramel-candy glass.

By the way, you can forget the word "crystal" that glassware manufacturers apply to their better-quality merchandise; scientifically speaking, it's just plain wrong. A "crystal chandelier" or "crystal goblet" is made of glass that is as amorphous as any other. It's just a particularly clear and brilliant quality of glass, usually containing lead oxide.

Okay, now it's time to address the urban legend that simply

will not die: that the windowpanes in several-hundred-year-old buildings are thicker at the bottom, having flowed down somewhat over the years as any good supercooled liquid should. If you examine old cathedral windows that still have their original glass in them, you're sure to find many that are indeed thicker at the bottom. The trouble is that nobody has ever done measurements on enough panes to determine if significantly more of them are thicker at the bottom than at the top, middle or sides.

But even if a significant fraction of old panes were found to be thickest at the bottom, it wouldn't prove that they had flowed. Early window glass was made by methods that were quite crude compared with our modern plate glass processes, and uneven thicknesses were tolerated as being preferable to more serious defects such as bubbles and scratches. Now if you were a workman assembling windows from panes of uneven thickness, wouldn't you be inclined to set them in with their thickest parts at the bottom?

You Didn't Ask, but . . .

There must be some relatively high temperature at which glass does begin to flow. What is that temperature?

Glass experts talk about a "transition temperature" at which rigid glass does indeed become slightly plastic. For ordinary window glass, the transition temperature is about 1000 degrees Fahrenheit (550 degrees Celsius). Everyone must agree that the windowpanes in old buildings never got that hot.

Equal-ibrium

If water freezes at exactly zero degrees Celsius, and ice melts at exactly zero degrees Celsius, what would happen to a bowl of ice and water at exactly zero degrees Celsius?

Absolutely nothing, as far as you would be able to tell. The ice and the water would dwell in peaceful coexistence. But down at the molecular level, a chaotic dance would be going on.

Zero degrees Celsius (32 degrees Fahrenheit) is indeed both the freezing point of liquid water and the melting point of solid ice. You are undoubtedly picturing a poor little zero-degree water molecule that can't make up its mind whether to flow or float, to be liquid or solid. Well, that's really a good way to look at it, because the individual molecules do indeed get to make decisions, in a manner of speaking.

Let's consider first what goes on when liquid molecules freeze.

There are some rather strong attractions between water molecules that tend to make them stick together. (In Techspeak, they are called *hydrogen bonds* and *dipole-dipole attractions*.) In liquid water, the molecules are moving fast enough that the attractions can't really take hold. But as water—or anything else for that matter—cools, its molecules move more and more slowly (see p. 79). Zero degrees Celsius happens to be the temperature at which the molecules are moving just slowly enough that they can grab hold of one another with their attractive forces and settle down into the unique fixed positions that characterize ice. Ice's molecules are tied rigidly in place; they can't go swimming around the way liquid molecules do.

Now let's put an ice cube into liquid water. Some of the ice molecules at the surface of the ice will break their attachments to their fellows and join their freely swimming brethren. In other words, they will melt. Meanwhile, some of the liquid molecules near the ice's surface may be moving more slowly than the capture speed (they're not all moving at the same speed), and they will freeze onto the ice. So both melting and freezing can be taking place simultaneously, some molecules going one way and some going the other.

As long as the water is slightly warmer than zero degrees Celsius, there will be more melting going on than freezing, because there won't be enough slow water molecules to be captured onto the ice. Conversely, if the water's temperature is slightly lower than zero degrees, there will be more freezing going on than melting, because there will be more slow water molecules to be captured. At exactly zero degrees, there will be just as many ice molecules melting as there are liquid molecules freezing. Millions of tiny molecules are going each way, but from our relatively gigantic human perspective, we see absolutely nothing going on. The ice and water just sit there—until, of course, they begin to warm up, and then melting takes over.

In a sense, then, zero degrees Celsius is neither the "melting point" nor the "freezing point" of water. It's the temperature at which melting and freezing are happening *equally*. Scientists call this exact-balance point an *equilibrium* point. They would say that at zero degrees Celsius, ice and liquid water are "in equilibrium."

Equilibrium is a very important concept in chemistry because there are many situations in which, down at the molecular level, two opposing processes are going on at equal rates, so that up here at the human level we see no apparent change.

For more, look up "equilibrium" in the index of any chemistry book. But I warn you: There may be lots of equations.

Melting and freezing are so closely interrelated that just by touching an ice cube you can kick some water molecules from liquid to solid.

TRY IT   Wet your fingers and touch the ice cubes in your freezer. The cubes may stick to your fingers so tightly that you can lift them up. The ice cooled the water on your fingers down to its own temperature, which is obviously below the freezing point. When the water on your fingers froze, it grabbed on to the ridges of your fingerprints and held on to them while simultaneously fusing itself onto the ice cubes, thereby "gluing" your fingers to the cubes.

Help! We're Trapped in an Ice Cube!

Why are my ice cubes cloudier in the middle than at the edges?

The cloudiness is a mass of tiny air bubbles—air that was dissolved in the water and expelled when the water froze. You can see the individual bubbles through a magnifying glass.

There is always some air dissolved in any water that has been exposed to . . . well, the air. For this, the world's fish are truly grateful. They are particularly grateful for the fact that, even though air is only about 21 percent oxygen, oxygen dissolves in water twice as readily as the other 79 percent, which is mostly nitrogen.

When water freezes, the loosely moving water molecules settle down into rigid positions. In doing so, they squeeze out the dissolved air molecules, because there is simply no room for them. When the water begins to freeze, the outer portions freeze first because they are in the best position to have the heat sucked out of them. As the dissolved air molecules are squeezed out, they become trapped within the encroaching casing of ice. The air molecules are forced closer and closer together as the growing wall of freezing water closes in on them. Eventually, they are packed so close together that they congregate into bubbles. And there they remain, trapped when the interior water finally freezes.

Help! I'm Breathing!

My girlfriend is worried that if the humidity gets to be 100 percent, we'll be breathing pure water and drown. Of course, that's silly, but I can't explain why.

Ask your girlfriend, "A hundred percent of what?"

Chicken Littles who fear drowning in air are forgetting

that "humidity" is purely relative. Everybody goes around talking about "the humidity" as if it's something absolute, but it's really the *relative* humidity that they're talking about—relative to some maximum, but still small, amount of water vapor in the air. And mind you, that's vapor, not liquid. Even when the relative humidity gets to be 100 percent at room temperature (we'll see that humidity varies with temperature), there is only one water vapor molecule in the air for every forty or fifty air molecules.

"Vapor" is a funny word. All it means is "gas"—the form of matter in which the molecules are flying freely around with huge spaces between them. Any substance can be transformed into a gas if we heat it hot enough to drive its molecules completely away from one another. It's only when the gas in question recently arose from a liquid that we refer to it as a vapor. We call the oxygen in the air a gas because we've (most of us) never seen it as a liquid. But we don't generally refer to gaseous water as a gas because we know that it came from liquid water. We call it "water vapor."

Why does water choose to go into the air as vapor, anyway? At every temperature, water finds a unique balance point between its tendency to exist in the form of liquid and its tendency to exist in the form of vapor (see p. 208). At warmer temperatures, the balance favors vapor, because the molecules are moving faster and can escape more easily. So the higher the temperature, the higher the tendency for water to be in the form of vapor, rather than liquid.

If you put some water of a certain temperature into a closed box, it will fill the box with the amount of vapor that is characteristic of its temperature, and then stop. It stops when there are as many liquid molecules leaving the liquid each second as there are vapor molecules hitting the liquid's surface and sticking. When these two rates are equal, there is no further net change. (Techspeak: The liquid and the vapor are *in equilibrium;* see p. 200.)

A lot of people, including some scientists, would like to say that the air in the box is now *saturated* with water vapor, as if the air were a wet rag, holding as much water as it can. But that's a misleading way to look at it. We'll put it another way: The amount of vapor in the box is 100 percent of the maximum amount that there can be at that temperature. In other words, the relative humidity is 100 percent. If there were only half that amount of water vapor, we would say that the relative humidity is 50 percent, and so on.

If we lived in a closed box with some liquid water in it, the relative humidity would always be 100 percent—100 percent of the maximum amount of vapor for that water temperature. But of course, we don't. We live in a constantly shifting sea of winds bearing warm air, cold air, high- and low-pressure air and everything else that the weather can devise to blow water vapor around from place to place. That's why the relative humidity isn't always 100 percent, even when it's raining, or even over the ocean.

Frighten your timid friend with this fact: In a steam bath or wet sauna the relative humidity is 100 percent and then some. First of all, the temperature is deliberately kept high to get as much water vapor into the air as possible. But in addition to that maximum amount of water vapor, there are actually tiny droplets of liquid water suspended in the air. We call it steam or fog. In a steam room, you're actually breathing liquid water. But nobody ever drowned from breathing fog or steam at a reasonable temperature because there is still plenty of air in between the suspended droplets.

(Caution: Steam can be dangerously hot, depending on how it has been produced and at what pressure. The steam in a steam bath is "cool steam" and is no hotter than the air in the room.)

You Didn't Ask, but . . .

What is the "dew point" that weather reporters are always talking about?

Meteorologists love to tell us the dew-point temperature, even though few of us know what it is and even fewer of us care. But as long as I'm this deep into water vapor, I might as well explain that too.

The dew point, or dew-point temperature, is the temperature below which the liquid-vapor balance of water shifts to favor the liquid side of the scales. That is, condensation wins out over evaporation.

If the temperature is above the dew point, liquid water will continue to change into vapor until it is all evaporated; wet things will dry. But if the temperature is below the dew point, the balance shifts in favor of liquid, and vapor will tend to condense. When that happens up in the atmosphere, the vapor condenses into masses of microscopic droplets of liquid water that are too small to fall and stay suspended in the air. We call these masses of tiny water droplets clouds.

A more earthbound example: If the ground gets any colder overnight than the dew-point temperature, water vapor in the air will condense out onto the grass and leaves as drops of liquid dew. That's important to farmers because dew is, after all, free water for the crops. Also, there are ecosystems in the world where it almost never rains and where small animals depend on dew for their water supply.

NITPICKER'S CORNER

I wish that people wouldn't use the word *moisture* to mean water vapor. "Moist" means slightly wet or damp, and moisture is the water—*liquid* water—that makes an object moist. Yet people use "moisture" to mean water *vapor,* which is water in its gaseous form. Air can't hold moisture, meaning liquid water; it can only be said to "hold" water vapor, but even that isn't really correct (see p. 208).

Cosmetic advertisements, especially for "moisturizing" skin creams and lotions, just love to use the word "moisture" when they mean "water." Although moisture is nothing but water, that's considered too common a word for such elegant products. So the next time you're in a fancy restaurant, be sure to ask the waiter for a glass of moisture.

And by the way, what do cosmetic "moisturizers" do, anyway? Do they add moisture, or water, or whatever you want to call it? No. If that were true, why couldn't you just put water on your dry skin? "Moisturizing" lotions and cosmetics coat your skin with a concoction of oils and other water-blocking chemicals, so that your skin's supply of natural water stays in instead of evaporating. It sounds paradoxical, but the cosmetic's oil produces water.

Fire Makes Drier

Why does a hair dryer have to both heat and blow?

This is one of those questions that seems so natural that we forget to ask them. But that's what I'm here for: to make you wonder about things you take for granted, and then to replace your complacency with the smugness of knowing.

The water in your hair or clothes must first be converted from liquid into vapor before it can be spirited off by a stream of air. Blowing away liquid water isn't easy, as you can tell by the hurricane-force winds they have to use to dry your car at the car wash. Warming the liquid water in your hair or clothes—and that's what the warm air does—speeds up the water's molecules, so that more of them can fly off into the air. (Techspeak: Warmer water has a higher *vapor pressure;* see p. 208.) The heat therefore speeds up evaporation of the water, and once it has evaporated into the form of vapor it can be swept away by the blowing air.

But how much vapor can the air-heated water produce? How fast can it evaporate?

Liquid water molecules can keep flying off and becoming vapor molecules only until the space above the liquid is so crowded with vapor molecules that just as many are bouncing back *into* the water as are flying *out* of it. (Techspeak: until the liquid and vapor are *in equilibrium.*) That's where the blowing comes in. The moving air from the dryer blows away some of those water vapor molecules so that they can't go back into the liquid. This "makes room" for more, and the evaporation continues.

That's why clothes and hair dryers do both heating and blowing. One without the other wouldn't do the job nearly as well. What if your hair dryer's blower blew out, so that it only heated your hair, or if its heater blew out and it blew cold air?

If there is already a lot of water vapor in the air—for example, if the bathroom is already steamed up and humid from your shower—the water in your hair won't be able to evaporate as fast. It will require a much longer heating and blowing time to dry your tresses to that silky, sexy, slow-motion slinkiness that they show on TV.

To Have, but Not to Hold

Why is it that warm air can hold more moisture than cold air can? That's why it's more humid in the summer, isn't it?

No. It's usually more humid in the summer because there's more water vapor available.

I don't mean that the oceans, lakes and rivers somehow expand in the heat. (Well, maybe a tad.) More precipitation? Perhaps. But it's not the amount of water itself; the humidity can be quite low over the middle of the ocean. What counts is how much water is being converted to *vapor* (or gas; see p. 203). It is more humid in the summer because the water supplies—the oceans, lakes, rivers and rains—are warmer, and water has a greater proclivity for making vapor when its temperature is higher.

Notice that I have said nothing at all about the air or its ability to "hold water." Humidity is the water vapor that comes from water, whether there is any air present or not. When we say, "It's humid today," we assume that the "it" in question is the air because, after all, what else is there? But the air plays no role whatsoever in humidity; like Mount Everest, it is simply "there." It is a bystander.

Think of it this way: We happen to be immersed in a sea of air, just as fish are immersed in a sea of water. If somebody suddenly dumps a load of red ink into the ocean a fish might say, "My, but it's red today." But "it" isn't the water itself; "it" is the ink that has been mixed into the water. Well, humidity is water that happens to be mixed into the air.

Nevertheless, you'll hear scientists and meteorologists explain humidity and other weather phenomena by talking about the "amount of moisture that the air can hold" and saying that warm air can "hold more moisture" than cold air can. That's a mistaken and misleading notion. The air isn't holding on to water vapor; it has no such holding power.

Here's why. Air and water vapor are both gases, and in gases the spaces between the molecules are so vast that any two gases can mix in any proportions without either one "knowing"—or controlling—how much of the other is there. All the air can do is accept the water vapor—whatever amount the water chooses to give off according to its temperature. *It is purely water's decision as to how eagerly it wants to be in the form of vapor instead of liquid.*

Now I suppose you want to know *why* warm water produces more water vapor than cold water does, right? That's science for you: Every answer generates more questions.

Water—like all liquids—has a certain tendency for its molecules to leave the surface of the liquid and fly off into the air. That's because all the molecules are moving with various speeds, and there will always be some of them at the surface of the liquid that have enough energy to go flying right off as vapor. Because molecules move faster on the average at higher temperatures than at lower temperatures, there will be more potential escapees from warm water than from cold. For example, at 86 degrees Fahrenheit (30 degrees Celsius) water produces seven times more vapor molecules in a given space than water at 32 degrees Fahrenheit (zero degrees Celsius).

There is always a sort of tug-of-war going on among the molecules of a liquid. Their speed wants them to fly off as vapor, but their attraction to their fellow molecules wants them to stay in the liquid. At every temperature, water must strike a balance between these two tendencies (Techspeak: The water attempts to come to *equilibrium* between these two states; see p. 200). At lower temperatures, the liquid tends to win out; at higher temperatures, the vapor gets the edge because of the higher molecular speeds. (The ultimate limit is when the liquid boils and turns completely to vapor.)

At a given temperature, every liquid has its own preferred balance point between vapor and liquid, because its molecules have their own degree of stickiness toward one another. A liquid whose molecules are stuck tightly together will not form vapor easily, so its balance point will tend to favor the liquid form over the vapor form. Gasoline's molecules, on the other hand, don't stick to one another very much at all, so their balance point favors the vapor and gasoline evaporates (vaporizes) much faster than water.

The tendency of liquid molecules to escape and fly off as vapor is called the *vapor pressure* of the liquid (see p. 207). In Techspeak, we would say that gasoline has a higher vapor pressure than water, and that warm water has a higher vapor pressure than cold water.

Let's say we're in a closed box containing some water. The water would soon strike a balance between liquid and vapor, according to the temperature. (The liquid and vapor would be *in equilibrium*.) If our box suddenly got colder, the water would have to strike a new balance (find a new equilibrium balance between liquid and vapor) based on that new, lower temperature. The new balance would be in the direction of less vapor and more liquid, so some of the vapor would have to condense out and become liquid. There would be rain or dew in our box.

Others may claim that it rained because "there was more water vapor in the cold air than it could hold." But I never even said there was air in the box, did I? It rained solely because the water shifted its liquid-vapor balance all by itself. "It" would be just as humid or just as dry in the box if there were no air in it at all, but some other gas with a different reputed "holding power."

I Can't See Where I'm Going!

When the windshield of my car gets all fogged up, how can I clear it most quickly?

Your car is a box of homemade weather, produced by your air intakes, your heater, your air conditioner and your passengers. But sometimes, because of your passengers' irrepressible habit of breathing, the car fills up with a lot of water vapor and some of it condenses onto the cold windshield, fogging it up. What do you do?

TRY IT When it's very humid in your car and the windshield fogs up with condensed moisture on the inside, turn on the air conditioner, no matter how cold it may be outside. (You can always turn on the heater, even while the AC is on.) Direct the conditioned air onto the windshield and it will clear up in a jiffy.

What happened was that the air conditioner took in the water vapor (along with the air that it's mixed with) and cooled it down to a lower temperature at which the water would much rather be liquid (see p. 208). It condensed into liquid at the AC, where it was thrown away outside the car. There was then not enough water vapor in the car for the temperature, and the liquid on the windshield restored the balance by turning itself into vapor. Voilà! A dry windshield.

But what about the rear window? When *it* fogs up, there's no way to blow air-conditioned air onto it; the cool air all comes out of ductwork up front, where the driver needs it, and there is no blower for the rear window. So what did those clever car designers do? They embedded heater wires in the rear glass. Instead of blowing dried-out, cool air on it, you just heat the glass. That raises the glass's temperature above the point at which water prefers to be liquid, so it turns to vapor and the fog disappears.

Odd, isn't it? To defog the windshield you cool the air, but to defog the rear window you heat the glass, and the end result is the same. Why don't the car manuals ever explain this to you? How many people are driving around with fogged-up glass, not having the foggiest notion of what to do about it?

Now how about your bathroom mirror back home? After your shower, it's fogged up worse than any windshield ever got in the steamiest jungle, and just when you want to shave or put on your makeup. I'll bet you have neither an air conditioner in your bathroom nor heating wires embedded in the mirror, so you can't use either of the car window tricks on it. But you probably have a hair dryer handy. Just sweep across the mirror with it, as if you were painting the glass with hot air. The dryer's air will heat the condensed water on the glass enough that it will prefer to be vapor rather than liquid and it will evaporate, just as it does on your electrically heated rear car window.

The Smell of Rain

A farmer neighbor tells me that he can smell when it's going to rain. Is he joshin' me?

Probably not. It's not the rain itself that he smells, but just about everything else. Almost everything smells a little stronger when it's about to rain.

Stormy weather is usually preceded by a drop in atmospheric pressure, or what the TV weather people like to call "barometric pressure." (Is that what you feel when you're struck by a falling barometer?) That is, before it rains the pressure exerted by the air drops, and it presses down less heavily on the countryside.

Meanwhile, all the trees, grass, flowers, crops and, yes, even livestock are emitting their characteristic odors. Odors are tiny amounts of vapors emitted by substances (see p. 143), and we smell them when the vapor molecules happen to migrate through the air to our noses. When the air pressure is low and isn't pressing down so hard, it allows more of these vapors to escape into the air, and everything smells a little bit stronger.

Also, when the rain-bearing low-pressure front moves in, it is accompanied by a wind that carries along distant smells that are not ordinarily detected.

Of course, a farmer gets to be pretty good at reading weather clues, so he might be cheating a bit by consulting the sky, the winds, and even his arthritis.

By the way, doctors used to believe that people with arthritic joints could feel rain coming on because there are tiny gas bubbles in the joints, and when the air pressure decreases, those bubbles expand and cause internal joint pressure. Nice theory, but I understand that it's no longer in vogue.

7

Stuff and Things

Ours is a materialistic society.

We may talk about the birds, the bees, the trees, the moon and the stars, but what we surround ourselves with is an accumulation of goods—stuff and things that have been manufactured, sold, bought, used and ultimately thrown away. Even when trekking through the wilderness we must have a sleeping bag, a canteen, a knife and, customarily, some clothing. (Mosquitoes can be fierce.) Manufactured goods all.

Science resides within everything. Every artifact has perfectly good reasons, often unsuspected and fascinating reasons, for being precisely what it is and nothing else. I'm not talking about its invention or the technology of its manufacture. Invention and technology are not science; they are applications of science. I'm talking about the fundamental principles that endow each substance or object with its unique individuality as a stuff or a thing.

In this chapter we'll examine the materials and articles that we use daily, or almost daily—everything from soap to soda pop, erasers to explosives, rubber to radioactivity, airplanes to automobiles, aluminum foil to skateboards.

And then we'll end this book of answers by tackling the **Most Fundamental Question in the World:** Why do some things happen and other things not happen? Believe it or not, there *is* a general answer.

Extra! Shark Bites Airplane

Every time I get jittery about flying, someone tells me that the chances of being in a fatal airplane crash are much smaller than the chances of being attacked by a shark. What I don't understand is how a lot of shark attacks make my airplane any safer.

Congratulations. You've put your finger on the most flagrant example of figures that lie, with the possible exception of breast implants.

Let's take a look at some statistics.

From 1994 to 1997, the average number of shark attacks per year in U.S. waters was thirty-three. In the same four years, the average number of fatal airplane crashes was three. So you're eleven times less likely to be in a fatal crash than to be attacked by a shark, right?

Wrong. It is utterly meaningless to compare two such completely different sets of circumstances. Why on earth would anyone want to mention shark attacks and plane crashes in the same breath, except to push a predetermined point of view? Anyone who is comforted by such a fake argument is more likely to die of gullibility than from either a shark or an airplane.

But even if it were relevant to compare such totally unrelated sets of figures, they would still be meaningless without a lot of other information. Did the people who were killed in air crashes fly a lot more, or a lot less, than you do? If so, their odds were different from yours. And how many of America's 250 million citizens even went into the water? Did they do it in Florida, where most shark attacks occur, or in New York, where all the dangerous animals are in the zoos and subways?

What people fail to understand is that when you're in an airplane, your chances of dying in a crash are infinitely *higher*—not lower—than your chances of being attacked by a shark, because except for lawyers, there are no predatory beasts on airplanes. Worry about sharks when you're in the water; worry about airplanes when you're in the air. The only possible connection would be if your plane crashes into shark-infested waters, in which case your statistical ass is up for grabs.

But a legitimate question remains: How *much* should you worry about airplanes when you're in the air? What are the *relevant* statistics?

I feel safe when flying because of one statistic only, and it has nothing to do with the numbers of deaths by shark attacks, drowning, accidental falls, suicides, auto accidents or lightning strikes—numbers that are frequently quoted to assuage the fears of white-knuckled passengers, and all equally irrelevant. The one statistic that I keep quoting to myself has to do with my particular flight, the one I'm actually on. And the probability of any one flight ending in a fatal crash is about 1 in 2.5 million. That's plenty good enough for me.

Except when my flight gets bumpy, of course.

Here's the Rub

How does an eraser erase pencil marks?

It doesn't work like a chalkboard eraser, which wipes an accumulation of chalk off a smooth surface. Paper isn't that smooth and a pencil mark isn't all on the surface; most of it is embedded among the paper fibers.

If you look at a pencil mark under a microscope, you'll see that it's not continuous; it is made up of individual black particles, a few ten-thousandths of an inch (several thousandths of a millimeter) in size, clinging to the paper fibers and tangled among them. The eraser's job is to pluck out these tiny particles. It can do that because (a) it is flexible enough to reach down between the fibers and (b) it is just sticky enough to grab on to the black particles and pull them out.

But while the eraser is rubbing the paper, the paper's fibers are rubbing off pieces of the rubber, which is now the rubbee, so to speak. The rubbed-off shreds of rubber roll up their collected black particles into those pesky crumbs that you have to brush away. Under the microscope, those crumbs look like enchiladas rolled in coal dust, to coin a rather unappetizing simile.

The black particles are made of graphite, a shiny, black mineral form of carbon that breaks apart easily into flakes. In its pure form graphite is much too crumbly to make sharp, detailed lines, so it is mixed with clay as a hardener and wax as a binder, in order to make pencil "leads" which, it behooves me to point out, contain no lead at all.

Real lead is a soft, gray metal that will leave dark marks when rubbed on a smooth surface. It was used for writing until the mid–sixteenth century, when a Swiss naturalist named Conrad von Gesner (1516–1565) put some graphite into a wooden holder and made the first pencil. Up until that time, graphite was thought to be a form of lead, and with an astounding display of inertia we are still calling the black stuff in pencils "lead" more than four centuries later. Even Herr Gesner was apparently unable to prevent the German word for his pencil invention from becoming *bleistift*, which means "peg of lead." And while we're at it, the English word "pencil" comes from the Latin *penicillus*, meaning a brush. And if you must know, *penicillus* is the diminutive of *penis*, meaning a tail. Don't blame me.

BAR BET   Lead pencils contain no lead.

So-called soft pencils, like the notorious No. 2s that we have all had to use on machine-scored tests, are not darker because the particles of graphite are any blacker. It's that more of them are deposited on the paper because of a higher ratio of soft graphite to hard clay in the pencil. The larger proportion of graphite allows more and bigger black particles to be scraped off onto the paper, which makes a broader and denser mark.

Graphite is a shiny mineral, and the dense marks that are left on the test sheets when we fill in the spaces with our No. 2 pencils reflect light. The sheets are run through a machine that scans them with light beams and searches for reflections. If light is reflected from the right places but not from the wrong ones, congratulations. You get a good grade.

And finally, did you ever wonder why rubber is called rubber? Because it's used to rub out pencil marks, of course.

In 1752, a member of the French Academy first suggested that coagulated caoutchouc, or latex, a gummy sap from certain South American trees that we turn into rubber, could be rubbed on lead marks to erase them. In 1770 the English chemist Joseph Priestley (1733–1804) was apparently so impressed that he named the substance "rubber," and it stuck. Today the name makes little sense, because rubber automobile tires, gaskets and so on aren't used to rub anything. Even the name "graphite" makes little sense today, because it was derived from the Greek word *graphein,* meaning "to write," and graphite has many other uses, in items ranging from lubricants to golf clubs.

All of which shows that writing has always been of more importance to mankind than automobiles or golf, although in our present society you'd never know it.

Understanding Rubber Is a Snap

Why does rubber stretch?

If there is one statement that will make you a passionate believer in molecules, it's this: Rubber stretches because it is made of stretchy molecules. A rubber band stretches because each of its molecules, all by itself, is built like a miniature rubber band.

Rubber molecules are shaped like long, skinny worms, all coiled and curled up, but capable of being straightened out by appropriate tugs on their heads and tails. A piece of rubber is like a can of these worms, all tangled together.

But think: You couldn't straighten out a whole snarl of worms by grabbing a head and a tail at random and tugging them apart; they would just slide past one another (unless they belonged to the same unfortunate critter). The two would have to be tied to each other in some way, so that a tug on one worm would get transmitted to its neighbor, and then to its neighbor's neighbor, and so on.

What we need is a can of worms that are spot-welded to one another in various places: Siamese worms, joined at the hip in various locations along their lengths. (Okay, so worms don't have hips, but you get the point.)

That's what the molecules of rubber are really like. But not at first, when the latex sap drips out of the rubber tree and is congealed and pressed into a sticky glob. Its molecules aren't spot-welded together very much, and especially when warmed they can slide easily over one another and the stuff becomes soft and gooey. A ball made out of raw rubber would hit the floor with a dull thud. And don't even think of making tires.

Humans therefore have to come along and accomplish the spot-welding themselves. They use a simple process called vulcanizing: heating the rubber together with sulfur. The sulfur atoms form bridges, or cross-links, between the rubber molecules, which allow them to stretch out to a certain extent, but unremittingly urge them back to their initial positions. That's why treated rubber is elastic: The molecules will stretch, but the cross-links will always bring them back.

Vulcanization makes wimpy, sticky raw rubber tough enough to be used in tires. The process was discovered in 1839 by Charles Goodyear (1800–1860)—yes, *that* Goodyear—who had been trying for ten years to find a way of making rubber tougher until he accidentally spilled some rubber mixed with sulfur onto a hot stove and it became tough and elastic. His discovery made him famous but not rich, and he died deep in debt.

That's the way the ball bounces.

TRY IT   Find a sturdy rubber band, at least a quarter of an inch (about half a centimeter) wide, and touch it briefly to your upper lip to test its temperature. Now stretch it quickly and while it's stretched, touch it again to your upper lip. It's warmer than it was.

What made it heat up?

When you stretched the rubber band, you put energy into it, didn't you? That energy warmed it up. The energy came from your muscles, and you'll have to eat a calorie or so of food to replenish it.

When a rubber band is stretched, its stretched-out molecules are in a more orderly arrangement—more lined up—

than they were in its unstretched, jumbled-up state. As everybody knows from doing housework, orderliness can be achieved only by putting energy into the job. So the stretched-out molecules must contain more energy—they're hotter—than the relaxed, unstretched molecules.

I have just sneaked in on you one consequence of the Second Law of Thermodynamics. This law of nature expresses the relation between energy and *entropy,* the degree of disorderliness of an arrangement. That's *dis*orderliness: the degree of random, haphazard scrambling. The Second Law recognizes that Nature's natural tendencies are (a) for energy to decrease—things tend to slow down and cool down—and (b) for entropy or disorderliness to increase—things tend to randomize and scatter. If you want to counteract the tendency for higher disorderliness (higher entropy), you have to increase the energy of the arrangement. That's an inevitable fact of nature; everything that happens is a trade-off between energy and entropy (see p. 248).

Life in the Contact Patch

What makes my new set of tires so much noisier than the old ones, especially when I drive fast? Sometimes I can hardly hear the police siren behind me.

I'll ignore the implications of that last part.

One obvious factor is that your old tires may have been pretty smooth, and smoother tires will be quieter.

Tire noise depends on the tread pattern, the roughness of the road and the roundness of the tires. Really, tires can be out-of-round, and the high spots will thump on the road during every revolution. But assuming that your tires are more round than square and that you're on a relatively smooth highway (that is, you're not in the state of Pennsylvania), the real question is why rolling rubber should make any noise at all; you'd think there would be nothing quieter. And indeed, tire manufacturers put a lot of effort into making their products operate as quietly as possible. Here are some of the things they have to consider.

As you can guess from the complexity of the sound—it's hardly a pure musical tone—it comes from a combination of several factors that make the tires vibrate. When the walls of a tire vibrate, it makes the air inside and outside vibrate also, and that's exactly what sound is: vibrations of air (see p. 28).

The source of most of the vibrations is the "contact patch"—the constantly changing flattened area that is in contact with the road. As each segment of the tire comes around in turn, it slaps against the road and is flattened into the contact patch. That constant slapping makes noise. But your new tires aren't perfectly smooth (unless you're a race driver; see p. 7). They have crosswise tread grooves that divide them into separate blocks of rubber, and those blocks hit the road in a rat-a-tat-tat machine-gun sequence. More noise. Moreover, as each block of rubber reaches the back end of the contact patch it snaps back into shape, again making the air around it vibrate. Still more noise.

A less obvious source of noise involves the escape of compressed air. As the tire turns, a groove entering the contact patch can trap some air and compress it against the road. Then when the groove leaves the contact patch the trapped air is released to the rear with a sudden—if you'll excuse me—fart. Experiments have been done with porous road surfaces that can significantly reduce this source of noise by allowing the air to bleed off directly into the road.

All of these effects depend on the groove pattern in your tires and the characteristics of the road surface. And the faster you go, of course, the more times per second all of these noise-producing processes are taking place. Driving at slower speeds will not only reduce the noise coming from your tires but will completely eliminate that annoying siren.

X-Rated Windows

A dirty film always seems to build up on the inside of my car's windshield. I don't smoke or allow anybody to smoke in my car. What makes that film?

One good film deserves another, so I'll borrow my answer from a famous line in the 1967 movie *The Graduate*: "One word: 'plastics.' " It's mainly the plastics in your car that produce the coating on your windshield.

Remember how your car smelled when it was brand-new? It *smelled* brand-new. A new car's smell is a potpourri of the many volatile chemicals used in its manufacture, from paint and cement solvents to chemicals used in treating rubber, plastics and fabrics and, if you're affluent, that "rich Corinthian leather" on the seats. In fact, every substance in the world is continually evaporating some of its molecules into the air to a greater or lesser degree. (Techspeak: Every substance has a certain *vapor pressure;* see p. 208.) We smell a substance when some of its evaporated molecules reach the olfactory nerve cells in our noses (see p. 143). And those that don't wind up in our noses might land anywhere else in the car.

Most of these volatile substances evaporate completely and dissipate long before you've paid off the loan, and when that happens your car no longer smells new. But other substances, smelling less noticeably of new debt, are released more slowly over a long period of time.

Plastics, in particular, are the biggest long-term emitters of chemicals: mainly plasticizers, which are waxy chemicals that give them flexibility. When your car is out in the sun, the intense radiation beating down through the windshield—modern automobiles have almost horizontal windshields for streamlining—hits the plastic dashboard cover and drives out plasticizer vapors, which then condense on the slightly cooler glass. The resulting, sticky film of waxy plasticizer then collects dust particles that blow in through the air ducts around the windshield, whereupon you may either clean it or buy a new car.

Wrinkle, Wrinkle, Little Bar

Why do my clothes get so wrinkled, and how do pressing and ironing get the wrinkles out?

Your clothes get wrinkled because you insist on putting them on a warm, moist, moving body. If you were cold, dry and motionless, you'd have no problem. No clothing problems, anyway.

It's heat and moisture that put the wrinkles in, and it's heat and moisture that are going to get them out. Dry, cold pressing, even with tons of pressure, will accomplish little; you need both the heat and the moisture that did the original damage.

It's hard to make generalizations about wrinkling and ironing, because there are so many different kinds of fibers that our clothes are made of these days. There are the synthetic (man-made) fibers, including nylon and a variety of polyesters and acrylics with various trade names. The synthetics are all chemicals known as polymers: materials that are made up of huge molecules, each of which consists of thousands of identical, smaller molecules, all strung together into enormously long (for a molecule) chains. Many of these synthetic fibers are sensitive to heat. That is, when heated they bend and when they cool they retain the bend. If that is done to a garment at the factory, it will keep its shape.

On the other hand, there are the natural fibers taken from plants, animals and insects, including cotton from the cotton plant, linen from flax, wool from various animals and silk from worms. I'll concentrate on one of the oldest and most wrinkle-prone of all the fibers: cotton. As far as wrinkling is concerned, it's one bad actor.

Cotton fibers are filaments of cellulose, a natural polymer that occurs in plant cells. (To split hairs, so to speak, a *fiber* is a single unit of cotton that is at least a hundred times as long as it is wide, while a *filament* is an extra-long fiber, a *strand* is made up of many filaments and a *thread* is made of many twisted-together filaments. I knew you'd want to know.)

A cotton fiber acts somewhat like a long, thin bar of metal, in that when bent slightly it will spring back to its original shape, but it can be bent only so far before it will stay bent. The amount of bending it can stand before retaining the crimp depends on the temperature. There is a certain temperature below which it will spring back and above which it will stay bent. For dry cotton, this so-called transition temperature is about 120 degrees Fahrenheit (49 degrees Celsius).

So far, you're lucky, because your body temperature is only around 99 degrees Fahrenheit (37 degrees Celsius), well below the crimping temperature.

But then there's the effect of moisture. Water, in the form of perspiration, for example, can lower the transition temperature of cotton down to around 70 degrees Fahrenheit (21 degrees Celsius). And whether you know it or not, you are always perspiring. You don't usually notice it because the perspiration evaporates from your skin as fast as it is produced—unless, of course, the air is very humid, in which case it doesn't evaporate and you say that you are "sweating."

So now if you actually have the audacity to sit on your pants or skirt, or to stress your other apparel by sharply bending your warm, moist limbs, the fibers can be bent into new, crooked shapes. Then when you stand up, your perspiration evaporates and the fibers cool down below their transition temperature and stay in those new shapes. Your clothes are wrinkled and you are rumpled.

How do you get the fibers back to their original, straight shapes? Just give them heat and moisture again, to get them above the transition temperature while you hold the fabric in its original flat shape with the sole plate of a steam iron. The steam lowers the transition temperature way below the temperature of the iron, so the original, straight shapes can reform.

An easy way to think of all this is that heat and moisture "melt" the structure of the fibers, and when they cool their shapes "freeze," whether those shapes happen to be straight or crooked.

In the laundry, always try to remove your clothes from the dryer while they are still warm and slightly damp. In that condition you can lay them out flat and they'll cool into that flat shape. Or you can hang them up and let gravity pull them flat. But if you leave them in the dryer too long after it stops, the clothes will cool down in their jumbled positions and the wrinkles will be set.

NITPICKER'S CORNER

Why is there a certain temperature above which cotton begins to wrinkle?

Cotton is made of cellulose, a natural polymer whose molecules consist of thousands of sugar (glucose) molecules, joined together into long chains. Cotton fibers are bundles of these cellulose molecules, all lying alongside one another in the direction of the fiber.

Here and there, the cellulose molecules are weakly bonded to one another sideways by so-called hydrogen bonds, which tie them together like a loose bundle of sticks. The trick is to keep them that way, because if those weak bonds are broken and re-formed while the fibers are bent, they will stay bent.

Hydrogen bonds can be broken by a combination of heat, which makes the molecules jiggle, and water, which swells the fibers by getting between the molecules. Water lowers the transition temperature because when the fibers are swollen the molecules are farther apart and easier to separate. That's why a steam iron works so much better than a dry one.

Ollie Oop!

I've seen kids do things on skateboards that seem to defy the laws of physics. As they jump over an obstacle the board rises with them, even though it's not attached in any way to their feet. How can that be?

What you're describing is a maneuver called an ollie, named after its inventor, Allen Ollie Gelfand. Gelfand was one of a number of southern California surfers in the late 1950s who just couldn't wait for good surf to come up and decided to surf the sidewalks. That's what started the skateboard craze.

An ollie is a jump into the air without the loss of one's skateboard. It does indeed look as if the board simply follows the feet, as if in a magician's levitation trick, and a good skateboarder does it so fast that you don't see how it's done. It depends on the fact that a skateboard isn't just a flat board on wheels—it has a bent-up tail at the rear end, and that's the secret to how it gets launched upward.

DON'T TRY IT	Learning to do tricks on skateboards takes lots of practice, not to mention antiseptics, bandages and splints. The following description may sound logical but is not intended to be a lesson.

Here's how a skateboarder does an ollie.

As he approaches an obstacle that he wants to jump over, the skateboarder places one foot in the middle of the board and the other one at the tip of the tail. He then stomps hard on the tail with his rear foot, which makes the tail hit the ground and the front end of the board (the nose) flip up like the opposite end of a seesaw. Simultaneously—and timing is critical—he jumps upward, hopefully high enough to clear the obstacle. As he becomes airborne, the board's nose will still be pressing upward against his front foot with momen-

tum that it received from the tail-stomp. He quickly slides his front foot forward to push the board's nose down level with the tail. He is now in midair on a level skateboard, sailing—again hopefully—with enough forward momentum to clear the far end of the obstacle (which, of course, requires that he had enough forward speed when he began the jump). Finally, as gravity begins to win out, he and the board fall together, with his feet still in contact with the board.

The important thing for the skateboarder to realize is that he can go no higher than he could by jumping straight up from a standing position. The amount of vertical travel he achieves is completely independent of any forward travel, because gravity doesn't know or care about any motion parallel to Earth's surface; it cares only about how far he is from Earth's center and whether it can pull him down any closer (see p. 104).

So if a skateboarder wants to sail over a picnic table, he must first make sure that he can jump straight up and onto the table before he tries it with a skateboard under him. And the board does add to the height that he must jump in order for it to clear the table along with him.

Note that the skateboarder and his board received their upward flight energies from two different sources: he from his leg-powered jump and the board from the tail-kick he gave it, which shot the nose into the air. (In fact, even without anyone on it, a skateboard on the ground would leap into the air when its bent-up tail is stomped on.) There's nothing magic, then, about the fact that board and rider go upward together, in spite of not being fastened to each other. An expert olliemeister allows no crack of light to show between his feet and the board, so it really does look as if they're glued together.

TRY IT   To see how a skateboard will fly upward from a stomp on its tail, place a spoon on the table, hollow side up. The turned-up bowl is like the turned-up tail of the skateboard. Now tap the end of the bowl sharply with a finger, as the skateboarder would stomp on his board's tail. The spoon goes flying upward, handle first, in a seesaw effect and then continues to flip end-over-end. If it were a skateboard, the rider's front foot would be holding down the handle end, shifting its momentum backward to the bowl end, which would then rise to the level of the handle.

Once a person masters the ollie and is released from the hospital, he can use it as the basis for any number of other skateboarders' tricks, all of which seem to involve life in midair. What tricks? How about a nollie, grind, heelflip, kickflip, ollieflip, pop shov-it, shov-it kick flip, casper, melloncollie, McTwist, tailslide, wheelslide, lipslide, indygrab or wallride? Many of these tricks are performed not on the street but in skateparks with artificial slopes, walls and slides in which competitions are held.

I'll stick to golf.

Soda . . . POP!

Why does the champagne gush out all over the place when I open the bottle? That stuff's expensive, and I hate to waste it.

You mean you actually want to drink it? Judging by what we see on television, you'd think that the major role of champagne in American culture is to hose down Super Bowl winners in locker rooms. Somewhat younger children do the same thing with soda pop, making sure to shake the bottle well before moving the thumb partially aside on the top in order to aim better. . . . Well, you know the rest. (DO NOT TRY THIS AT HOME!)

If I said that shaking a bottle of champagne, beer or pop raises the gas pressure inside, ninety-nine out of a hundred people, even chemists and physicists, would agree. But it's not true. When you shake an unopened bottle or can of carbonated beverage the pressure inside does not change.

It certainly does *seem* as if the pressure is increased by shaking, and it's easy to dream up smug theories as to why that should be. But I won't muddy the waters by quoting those theories here, because they've turned out to be all wet.

Then why does the liquid squirt out with so much force when you open a shaken bottle? It's only because shaking makes it easier for gas to escape from the liquid, and in its eagerness to escape when the bottle is opened it carries some liquid along with it.

It was two chemists named David W. Deamer and Benjamin K. Selinger at the Australian National University in Canberra who in 1988 settled the question in the simplest possible way: by measuring the gas pressure inside a bottle of pop before and after shaking it. They adapted a standard pressure gauge, not too different from a tire gauge, so that it could be screwed onto the top of a soda bottle.

Their results (which would have been the same if they had splurged and used champagne): If an unopened bottle has been standing quietly at room temperature for a day or so and is then shaken, the pressure of carbon dioxide gas in the head space (the space above the liquid) does not change.

The reason is that the gas pressure is determined by only two things: (a) the temperature and (b) how much carbon dioxide can dissolve in the liquid at that temperature (Techspeak: the *solubility* of the gas in the liquid). There is only so much carbon dioxide gas in the bottle; some of it is dissolved in the liquid and some of it is loose in the head space. When an unopened bottle of soda has remained at the same temperature for some time, the amount of gas dissolved in the liquid—and more important, the amount of gas that is *not* dissolved in the liquid—settles down to whatever the appropriate proportions are for that particular temperature. (Techspeak: The system comes to *equilibrium;* see p. 200.) You can't change those proportions by doing anything short of changing the temperature or adding more carbon dioxide.

(If you put the bottle in the fridge for twenty-four hours or so, more of the gas will dissolve in the liquid, because gases dissolve to a greater extent in colder liquids. There will then be less gas in the head space, and the pressure will be less. That's why you get less of an outburst of gas when opening a cold bottle than when opening a warm one.)

The point is that shaking alone can't change the pressure because it doesn't change the temperature or in any other way change the amount of force or energy that is available inside the bottle. So never fear that manhandling your beer, soda or champagne on the way home from the store will make the bottles explode. On the other hand, make sure not to let the bottles heat up in the trunk of your car, because the higher temperature will indeed raise the pressure of the gas.

Now we can take a more educated look at what causes the explosive emission when we open a recently shaken bottle. It is caused by an increase in the amount of gas that is set loose—not by heating, but by the mechanical "outing" of some dissolved carbon dioxide from the liquid when the bottle is opened.

Here's how.

First of all, a bunch of dissolved carbon dioxide molecules can't just decide to gather together in one spot and form a bubble. They need something to gather upon—a microscopic speck of dust or even a microscopic irregularity on the surface of the container. These congregation spots are called *nucleation sites,* because they serve as the nuclei, or cores, of the bubbles. Once a small gang of carbon dioxide molecules has gathered at a nucleation site and formed the beginnings of a bubble, it is easier for more carbon dioxide molecules to join up, and the bubble grows. The bigger the bubble gets, the easier it is for even more molecules to find it and the faster it grows.

Now when you shake a closed bottle of pop, you're making millions of tiny bubbles of gas from the head space that become trapped in the liquid. There, they serve as millions of nucleation sites upon which millions of brand-new bubbles can grow. If the bottle is then left to stand for a long time, the new baby bubbles will be reabsorbed and all the contents will return to normal, in which condition it is no longer a threat.

But those new nucleation sites and their newly hatched bubbles don't disappear very quickly; they remain for some time in a recently shaken bottle, just waiting for some unsuspecting soul to come along and open it. When he does, and the pressure in the head space suddenly drops to atmospheric pressure, the millions of baby bubbles are free to grow, and the bigger they get the faster they grow. The large volume of released gas erupts abruptly into a gigantic blurp that carries liquid out of the bottle.

BAR BET   Shaking a bottle or can of beer or soda pop does not increase the pressure inside.

Oh, the champagne? Same thing. The best way to handle it is to leave it undisturbed in the refrigerator long enough for it to "come to equilibrium"—at least twenty-four hours. Then be careful not to either warm or agitate it before or during opening. After removing the wire twist, ease the cork upward with your thumbs. All of the champagne will stay in the bottle and the cork won't become a lethal missile.

Diet Coke Loses Weight

My buddy claims that he can tell a can of diet Coke from a can of Classic (regular) Coke without opening them or reading the labels. Can he?

Probably. It isn't difficult, and it works with Pepsi too. It's based on the fact that a can of the diet drink is slightly lighter than a can of the regular drink.

Regular Coca-Cola is sweetened with sugar (sucrose) or corn sweeteners, which are other sugars, usually fructose, maltose and/or glucose. Diet Coke, on the other hand, is sweetened with aspartame, an artificial sweetener. Gram for gram, aspartame is 150 to 200 times sweeter than sucrose, so only a tiny amount of it is needed to produce the same sweetness as in the sugared product. While the amount of sugar in the regular drink is 2 or 3 percent, there are only a few hundredths of a percent of aspartame in the diet drink. Therefore, a can of the diet drink is very slightly lighter in weight.

Your buddy can't tell the difference just by hefting the two cans. But if he fills a sink with water and places the unopened cans in it, the diet can will float higher in the water than the regular can, which might even sink.

BAR BET   I can tell a can of diet Coke from a can of regular Coke without opening them or reading their labels.

Sour Power

In a novelty catalog I saw a "fruit-powered clock." You stick two wires into an orange or lemon, and it runs a small digital clock on "the natural energy found inside a fresh fruit or vegetable." What's the scoop?

"Natural energy" is a favorite buzz-phrase of hucksters and kooks pushing everything from arthritis cures to communication with the dead. There seems to be this idea that "natural energy" is everywhere, to be plucked out of the air by such magic trinkets as copper bracelets (for arthritis) or by those crystal amulets that you wear around your neck or fondle in your pocket to ward off what supposedly less sophisticated societies would call "evil spirits." If any of these things provided one-thousandth of the energy that their boosters expend in peddling them, we'd never have to burn coal or petroleum again.

As far as fruits and vegetables are concerned, their only "natural energy" is in the form of the calories that you get by eating them—the energy that you gain when you metabolize, or "burn," the food, just as you can release energy by burning a piece of coal. Eating coal, however, doesn't work because our bodies have no mechanism for digesting and metabolizing it—that is, for extracting its chemical energy.

Oranges and lemons contain precious little food energy, as you might guess from the fact that they don't burn worth a damn (except for the oils in the rind). Even if you could convert all its nutritional energy into electricity instead of muscle power, the fifteen calories in a lemon would keep a $7\frac{1}{2}$-watt night-light burning for only about two hours.

Other than that, the only way to get useful energy out of a lemon would be to drop it from a tall building.

Does the fruit clock actually work? Amazingly, it does. It will run for weeks or months with its wires thrust into a fruit or vegetable—almost any fruit or vegetable. "Potato-powered" clocks are quite popular, presumably because there's nothing quite so dumb and lifeless as a potato, and getting energy out of it appeals to people's sense of the ridiculous.

Here's how the veggie clocks work.

The wires that you thrust into the fruit are made of two different metals, usually copper and zinc. Together with the fruit juices in between, these two metals make a genuine electric battery (more properly called a voltaic cell, but we'll call it what everybody else does). All it takes to make a battery is two different metals with some sort of electricity-conducting liquid in between.

You know that an electric current is a flow of electrons going from one place to another—through a wire, through a lightbulb, through a motor or in this case through an electronic digital clock. The question is, How do you entice electrons into traveling from one place to another so they can run a clock along the way?

A battery induces electrons to travel because it contains two different kinds of atoms that hold on to their electrons with different degrees of tightness. For example, copper atoms hug their electrons more tightly than zinc atoms do. So if you give zinc's electrons a chance, they'll leave home and migrate to the copper, where they feel more wanted.

Clever humans that we are, we offer the electrons only one route from the zinc to the copper: through our digital clock. If they want to get to the copper, they'll simply have to force their way through our clock, operating it as they go.

Then why is the fruit or vegetable necessary? The juice inside it is what chemists call an electrolyte: a liquid that conducts electricity. It completes the circuit of electrons, restoring them and their charges to the zinc, which would otherwise quickly become so depleted of electrons that the whole process would stop.

So where does the "natural energy" actually come from? It's inherent in the constitution of the zinc and copper atoms—in their natural difference of electron-holding powers.

A battery is so easy to make that at least one may have been built by the Parthians, a people who lived two thousand years ago in what is now Iraq. In 1938 a German archaeologist described a small clay jar from that period, then in the National Museum in Baghdad. The jar contained an iron rod inside a copper cylinder; one needed only to fill it with fruit juice (or wine) for it to have enough kick to power an ancient Parthian digital wristwatch.

Okay, so nobody really knows what it was used for.

If indeed it was a battery.

If it wasn't a hoax.

If . . .

You Didn't Ask, but . . .

Why does the "Two-Potato Clock" need two potatoes?

For the same reason that your flashlight needs two batteries.

A set of zinc and copper metals will move electrons with only so much oomph. That's because there's only a certain amount of difference between the electron-holding powers of zinc and copper. But if you need more electron-moving force—to light a bulb, for example—you can connect a second set of zinc and copper metals after the first, giving twice as much kick to the electrons.

The technical word for electron kick is *voltage:* the force with which the electrons are made to move. The zinc-copper combination makes about 1 volt of kick. If a particular clock needs 2 volts to run, you'll need two potato batteries connected together.

There Are No Smoke Alarms in Hell

While changing the battery in my smoke alarm I decided to read the fine print on the label. It says that it contains radioactive material: americium-241. What does radioactivity have to do with detecting smoke?

What you have is an ionization-type smoke detector. It detects smoke by the fact that smoke interferes with air's ability to conduct a tiny electric current.

Under ordinary conditions, air doesn't conduct electricity at all; it's an excellent insulator. That's because the nitrogen and oxygen molecules in the air have no electric charge of their own, nor do they contain any loose electrons that could carry charge from one place to another, as metals do (see p. 97). If that weren't the case, electricity from those high-tension power lines overhead would zap right through the air to the ground, passing through anything—including us—in its way (see p. 100).

Air molecules—nitrogen, oxygen and a few others—don't have any net electric charge because the atoms of which they are made contain equal numbers of positive and negative charges that cancel each other out. The positive charges reside in the atoms' nuclei and the negative charges are in the form of electrons orbiting around the nuclei. But radioactivity can make air into an electrical conductor by knocking electrons out of the molecules, leaving them with some uncanceled positive charge. These electron-shy, charged molecules are called *ions,* and we say that the radioactivity has *ionized* the air. Because ionized air contains electrically charged molecules, it will conduct electricity.

How does radioactivity ionize the air?

The nuclei of radioactive atoms are unstable, and they spontaneously disintegrate by shooting out some of the particles of which they are made at speeds close to the speed of light. The nuclei of americium-241 choose to shoot out *alpha particles,* which compared with other radioactively emitted particles are as a baseball is to a BB. A hefty alpha particle can do a lot of damage to an atom that it hits, so it is very good at ionizing air molecules.

A tiny amount of americium-241 is packaged inside your smoke detector and its alpha particles keep a small region of air around it continually ionized. The battery provides a very small electric current that flows through that air. But when some smoke particles get into that air, the ions can collide with them and lose their charge. Less charge in the air means that less current can flow. A circuit detects this drop in current and triggers an ear-piercing alarm.

The amount of radioactive americium-241 in a smoke alarm is extremely small: usually nine-tenths of a microcurie, which corresponds to a quarter of a microgram. Even though that quarter of a microgram is emitting more than 30,000 alpha particles every second, they're nothing to worry about, because alpha particles are such weaklings at penetrating matter that they can be stopped by a sheet of paper. No alpha-particle radiation whatsoever gets out of the smoke alarm box.

NITPICKER'S CORNER

Whenever an atom of americium-241 (or any radioactive material) disintegrates, it is no longer the same kind of atom and doesn't have the same radioactive properties. So as time goes by, the remaining radioactive atoms decrease in number and so, therefore, does the amount of radiation they emit. In the case of americium-241, its number of atoms decreases by half every 433 years. (Techspeak: Its *half-life* is 433 years.) So 433 years from now, the americium-241 in your smoke alarm will be emitting only about 15,000 alpha particles per second. But don't throw it away yet, because after another 433 years it will still be working fairly well while emitting only 7,500 alpha particles per second. I'd advise you to replace it around 433 years after that, however, because by the year 3300 the electric current will be getting pretty weak and the alarm might go off even without any smoke. And those alarms, you know, can make enough noise to wake the dead.

Of course, if by then you're where I expect to be, smoke alarms aren't permitted because they'd be going off all the time.

Fertilizer Grow Boom!

Newspaper accounts of terrorist bombings have said that a chemical fertilizer was used as an explosive. How can a single chemical have such Jekyll and Hyde uses?

It's one of those coincidences that aren't really accidental when you dig a little deeper. As we'll see, the good-guy and bad-guy properties both stem from the fact that nitrogen gas is made up of molecules that strongly resist being torn apart.

First, the fertilizer role.

Every gardener knows that nitrogen is one of the three main elements that fertilizers provide, along with phosphorus and potassium. Nitrogen is extremely abundant; it makes up about 78 percent of the air we breathe. Its molecules consist of pairs of nitrogen atoms bound together into two-atom molecules, which chemists symbolize as N_2.

Those two nitrogen atoms are tied together so tightly that plants can't split them apart to get their nitrogen fixes. They need the help of lightning, which undeniably has enough power to do the job as it cracks through the air. Also, there are certain so-called nitrogen-fixing bacteria and algae that can split nitrogen molecules, but they haven't told us exactly how they do it.

We humans must resort to our powerful chemical technology in order to convert those nitrogen molecules into more plant-usable forms, such as ammonium compounds or nitrate compounds. The fertilizer ammonium nitrate contains nitrogen atoms in both of these forms, which makes it a doubly potent fertilizer.

Now what if the two separated nitrogen atoms in ammonium nitrate were suddenly given the chance to pair up again into strong molecules of nitrogen gas? They would grab that opportunity eagerly. After all, if nitrogen atoms love one another so much that when paired up they strongly resist being split apart, wouldn't they want to break out of the ammonium nitrate to reestablish their tight pairings and become nitrogen gas again? They would do that with such eagerness that they would literally explode out of the ammonium nitrate to rejoin each other and fly away into the air in blissful gaseous freedom.

I have just described an explosion: anytime a solid turns into a gas with great suddenness. The wave of released gases, which are expanding rapidly because of the heat that is also being released, is the pressure that does all the damage.

In the case of ammonium nitrate, which contains oxygen and hydrogen atoms as well as nitrogen, it's not just the nitrogen atoms that combine suddenly into tight gas molecules. Oxygen and water molecules are almost as tightly held together as nitrogen molecules are, so the oxygen atoms pair up into oxygen gas (O_2), while the hydrogen and oxygen atoms join up to form water vapor (H_2O). If given the chance, then, solid ammonium nitrate will suddenly break up and turn into an enormous volume of gases: nitrogen, oxygen and water vapor.

All it takes for ammonium nitrate to decompose violently in this way is heat: enough to reach a temperature of at least 570 degrees Fahrenheit (300 degrees Celsius). Even at temperatures as low as 340 degrees Fahrenheit (170 degrees Celsius), ammonium nitrate can explode, turning somewhat less violently into nitrous oxide gas and water vapor.

Keep your powder dry, certainly. But also keep your fertilizer cool.

Foiled Again

Why is one side of my aluminum foil shinier than the other?

It's because of a time- and space-saving shortcut that's used in the final stage of the manufacturing process.

Aluminum, like all metals, is *malleable;* that is, it will squish when enough pressure is applied. That's in distinction to most other solid materials, which will crack under pressure. So metals can be rolled out into extremely thin sheets.

Metals are malleable because their atoms are held together by a moveable sea of commonly owned electrons, rather than by rigid bonding forces between the electrons of one atom and the electrons of the next, as is the case in most other solids. In effect, then, it doesn't matter much where a metal's atoms are with respect to one another, and they are therefore free to be pushed around within the electron sea.

In the aluminum foil factory they roll sheets of aluminum through pairs of steel rollers that get progressively closer together, which squeezes the aluminum down to progressively thinner sheets. Household aluminum foil is less than a thousandth of an inch (two-hundredths of a millimeter) thick.

To save space in the final rolling, they feed a sandwich of two sheets at a time through the rollers. The top and bottom surfaces are in contact with the polished steel rollers and come out nice and shiny. But the inner surfaces of the sandwich are pressed against each other—aluminum against aluminum. Because aluminum is so much softer than steel, these surfaces press into each other somewhat, leaving a rougher, duller surface when they're separated. It makes no difference whatsoever in how you're able to use the foil.

And by the way: I hope you're not one of those people who sometimes call it "tinfoil." A foil is a very thin sheet of metal—any metal. Aluminum foil is a thin sheet of (surprise!) aluminum metal and tinfoil is a thin sheet of an entirely different metal: tin. Tin is a rather heavy, nontoxic metal whose foils were used as food and medicine wrappers before aluminum became cheap and widely available. But habits die hard, and many people still call aluminum foil tinfoil.

Someone should also let it be known that "tin cans" aren't tin, either. A "tin can" used to be a steel can lined with relatively noncorroding tin on the inside. But these days the linings of steel and aluminum cans aren't even tin; they're plastic or enameled coatings.

Avast, Ye Slob . . . uh, Swabs!

While sailing on a friend's boat, I didn't want to use up fresh drinking water, so I tried to wash my shirt in seawater. But I couldn't get any lather at all. Why doesn't soap work in salt water?

It's one of life's little ironies. Sailors do hard, often dirty work, yet with all that water around they can't bathe or wash their clothes with soap. Not with ordinary soap, anyway. There is a special soap called "sailors' soap" that works in salt water. But first let's see why the ordinary stuff doesn't.

It will not surprise you to learn that seawater contains a lot of salt—sodium chloride. Averaged over the world's oceans, every quart (liter) of seawater contains more than half a tablespoon (10 grams) of sodium chloride. It's the sodium that messes up the soap, because soap must dissolve in water before it can do its job and it won't dissolve well in water that contains a lot of sodium.

Soap molecules are made of sodium atoms attached to

long tails of what are known as fatty acids. The way soap works is that its fatty tail grabs on to the oily or greasy part of the dirt, while its sodium end drags it into the water. But if there are already too many sodium atoms in the water, the entry of still more of them in the form of soap molecules is inhibited. (In Techspeak, chemists refer to this situation as the *common ion effect,* because the sodium atoms, which are common to both the salt and the soap, are actually present as *ions,* or electrically charged atoms.)

This means that a sodium-containing soap won't dissolve enough in salt water to do its job of dragging sticky oil off the sailor and into the water, where it can be rinsed away.

But soaps don't have to be made with sodium. Potassium is a very close chemical relative of sodium's, and it too can combine with long fatty acid tails to make soap molecules. Compared with sodium, there is very little potassium in seawater, so potassium soaps aren't inhibited from dissolving. So-called "sailors' soap" is a potassium-based soap.

From Dust to Dust

Housecleaning is a never-ending round of dusting, dusting, dusting. If I stopped dusting, would my house eventually fill up floor to ceiling with dust?

You think *you've* got trouble? In China there are 2-million-year-old accumulations of dust (called *loess* by geologists) that are more than 1,000 feet (300 meters) thick. But it's not due to sloppy housekeeping. The dust has been swept up by winds from the Gobi Desert. In certain locations where the winds die down, they drop their loads of dust particles. The resulting huge dust piles became compressed from their own weight over the years, and some of them have actually been hollowed out into cave dwellings.

But never fear. At the rate at which the dust has accumulated in the Chinese loess cliffs, you could stop dusting in your house for a hundred years and still have a layer that is no more than an inch (2 centimeters) thick.

Unless you live near the Gobi Desert, you may be wondering where all the dust in your house comes from.

The dust in our atmosphere has many sources. Winds blow over dry earth, such as plowed fields, dirt roads and deserts. Plants give off pollen and other particulate matter. Forest fires and volcanos can spew dust and smoke particles high into the upper atmosphere, where they may blow around for years before settling. There is less dust over the oceans than over land, but still there are tiny bits of dried salt spray and even ash particles falling from meteorites that burn up in the atmosphere.

And, you're thinking, it all winds up on your bookshelves, right? Well, we're not done yet. Let's take a close look at the household dust that you generate yourself.

Notice that dust settles only on horizontal surfaces such as sills, shelves and the top edges of picture frames. (Forgot about those, didn't you?) Therefore, the stuff must be falling out of the air under the influence of gravity. That means that the particles of dust must be bigger than a certain size; if they were any smaller, the constant, agitated motion of the air molecules would keep them permanently suspended. That's the case with cigarette smoke, for example—the individual particles are so small that the bombardment of air molecules keeps them from falling. On the other hand, if dust particles were too big they wouldn't have been wafted into the air in the first place, later to come to rest upon that ugly china figurine that Aunt Sophie gave you.

But it's not all a matter of particle size. A relatively big tuft of lint from your clothing will float on the air because of its feathery shape, and this too will eventually find a landing pad someplace where you'd rather not have it. Those dust bunnies that take refuge in the windless climate under your bed are made up largely of fibers from clothing and other fabrics, often tangled together with fallen human or pet hairs and flakes of skin. (I never said this would be pretty.)

Everything that moves inside your house has the potential for sending out microscopic bits that are worn off and carried into the air. When a high-traffic area on your carpet wears out, where do you think all those carpet fibers went? Mote by mote, they wound up scattered around the house, to be dealt with on cleaning day.

Which brings up the question of how effective dusting really is. It depends a lot on how you do it. A dry dust cloth might just redistribute the dust, moving it perhaps from the shelf to the floor, demonstrating "to dust shall it return" in the literal, rather than the biblical, sense. Rubbing with a dust cloth can actually be counterproductive, because it can produce an electrostatic charge on the dust particles (see p. 188). Once charged, they can adhere tenaciously to any nearby object, so they will simply have been transferred from one object to another.

A dust cloth with a downy nap that traps the dust particles is one good idea. Another is to use one of those commercial dusting sprays. They contain an oil that not only makes the dust particles stick to the cloth, but coats them with a thin insulating layer so they can't adhere electrostatically to nearby objects.

For a surprising glimpse of how much dust there actually is in the air, look up at the beam of light coming from the projector the next time you go to the movies. The reason you can see the beam at all is that the light is being scattered sideways by ordinarily invisible dust particles that are approximately the same size as the light's wavelength (see p. 170).

To Go, or Not to Go?
That Is the Question

This may be a stupid question, but what makes things happen or not happen? I mean, water will flow downhill, but not up. Sugar will dissolve in my coffee, but if I put in too much I can't undissolve it. I can burn a match, but I can't unburn it. Is there some cosmic rule that determines what can happen and what can't?[3]

There's no such thing as a stupid question. You have asked what is perhaps the most profound question in all of science. Nevertheless, it does have a fairly simple answer ever since a genius by the name of Josiah Willard Gibbs (1839–1903) figured it all out in the late nineteenth century.

The answer is that everywhere in Nature there is a balance between two fundamental qualities: *energy,* which you probably know something about, and *entropy,* which you probably don't (but soon will). It is this balance alone that determines whether or not something can happen.

Certain physical and chemical things can happen all by themselves, but they can't happen in the opposite direction unless they get some outside help. For example, we could make water go uphill by hauling it or pumping it up. And if we really wanted to, we could get that sugar back out of the coffee by evaporating the water and then chemically separating the sugar from the coffee solids. Unburning a match is quite a bit tougher, but given enough time and equipment, a small army of chemists could probably reconstruct the match out of all the ash, smoke and gases.

The point is that in each of these cases a good deal of meddling—energy input from outside—is required. Left en-

---

[3]This question is so fundamental to the way in which the universe works that I'm repeating the answer, slightly modified, that I gave in *What Einstein Didn't Know: Scientific Answers to Everyday Questions* (Dell 1999). All other questions and answers in the present book are new.

tirely to herself, Mother Nature allows many things to happen spontaneously, all by themselves. But other things will *never* happen spontaneously, even if we wait, hands-off, until doomsday. Nature's grand bottom line is that if the balance between energy and entropy is proper, it will happen; if it isn't, it won't.

Let's take energy first. Then we'll explain entropy.

In general, everything will try to decrease its energy if it can. At a waterfall, the water gets rid of its pent-up gravitational energy by falling down into a pool. (We can make that cast-off energy turn a waterwheel for us on the way down, if we like.) But once the water gets down to the pool, it is "energy-dead," at least gravitationally speaking; it can't get back up to the top.

A lot of chemical reactions will happen for a similar reason: The chemicals are getting rid of their stored-up chemical energy by spontaneously transforming themselves into different chemicals that have less chemical energy. (The burning match is one example.) But they can't get back up to their original energy conditions by themselves.

Thus, other things being equal, Nature's inclination is that *everything will lower its energy if it can.* That's rule number one.

But *decreasing energy* is only half the story of what makes things happen. The other half is *increasing entropy.* Entropy is just a fancy word for *disorder,* or randomness, the chaotic, irregular arrangement of things. At the scrimmage, football players are lined up in an orderly arrangement—they are not disorderly, and they therefore have low entropy. After the play, however, they may be scattered all over the field in a very disorderly, higher-entropy arrangement. It's the same for the individual particles that make up all substances: the atoms and molecules. At any given time, they can be either in an orderly arrangement, or in a highly disordered jumble, or in any kind of arrangement in between. That is, they can have various amounts of entropy, from low to high.

Other things (namely, energy) being equal, Nature's inclination is that everything tends to become more and more disorderly—that is, *everything will increase its entropy if it can.* That's rule number two. There can be an "unnatural" increase in energy as long as there is a more-than-compensating increase in entropy. Or, there can be an "unnatural" decrease in entropy as long as there is a more-than-compensating decrease in energy. Got it?

So the question of whether or not a happening can occur in Nature spontaneously—without any interference from outside—is purely a question of balance between the energy and entropy rules.

The waterfall? That happens because there is a big drop in (gravitational) energy; there is virtually no entropy difference between the water molecules at the top and those at the bottom. It's an energy-driven process.

The sugar in the coffee? It dissolves primarily because there's a big entropy increase; sugar molecules swimming around in coffee are much more disorderly than when they were tied neatly together in the sugar crystals. Meanwhile, there is virtually no energy difference between the solid sugar and the dissolved sugar. (The coffee doesn't get hotter or colder when the sugar dissolves, does it?) It's an entropy-driven process.

The burning match? Obviously, there's a big energy decrease, a sudden exodus of energy. The stored-up chemical energy in the match head is released as a burst of heat and light. But there is also a huge entropy increase; the billowing flame, smoke and gases are much more disorderly than the compact match head was. So this reaction is doubly blessed by Nature's rules, being driven by both energy and entropy. That's why it proceeds with such gusto the instant you supply the initiating scratch.

What if we have a process in which one of the quantities, energy or entropy, goes the "wrong way"? Well, the process can still occur if the other quantity is going the "right way" strongly enough to overcome it. That is, energy can increase as long as there is a big enough entropy increase to counterbalance it; and entropy can decrease as long as there is a big enough energy decrease to counterbalance it.

What J. Willard Gibbs did was to devise an equation for this energy-entropy balance. If the Gibbs equation shows that after counteracting any "wrong-way" entropy changes there is still some energy left over, that energy (Techspeak: the *free energy*) can be used to make things happen and the process in question will take place automatically. If, on the other hand, the amount of available ("free") energy is inadequate to counteract any "wrong-way" entropy changes, the process will not and cannot take place unless some additional energy is obtained from outside.

By adding enough energy, then, we can always overpower nature's entropy rule that everything tends toward disorderliness.

Here's an example. There are about 10 million tons— 60 trillion dollars' worth—of dissolved gold distributed throughout Earth's oceans, just sitting there for the taking. With enough effort we could collect it all, atom by atom. But the atoms are dispersed through 336 million cubic miles (1.4 billion cubic kilometers) of ocean in a completely chaotic arrangement that has extraordinarily high entropy. The energy that we would have to expend in order to reduce its entropy by collecting it all in one place would cost enormously more than the value of the gold.

In a fit of fervor over the laws of mechanics, Archimedes (287–212 B.C.) is reputed to have said, "Give me a lever long enough and a place to stand, and I will move the world." If he had known about entropy and apple pie, he might have added, "Give me enough energy and I'll put this chaotic world into apple-pie order."

Some Techspeak Buzzwords

Words that are separately defined appear in *italics*.

Acceleration: Any change in the speed or direction of a moving object. It can be a speeding up, a slowing down or any deviation from a straight line.

Atom: A "building block" out of which all substances are made. Every atom consists of an extremely tiny and extremely heavy nucleus surrounded by a number of "whirling" *electrons*. There are over a hundred different kinds of atoms, distinguished from one another by the different numbers of electrons they contain. Atoms join together in various combinations to form a vast number of different *molecules*, making a vast number of different substances with different properties.

Density: A measure of how heavy a given bulk of a substance is. A liter of water weighs 1 kilogram. The density of water is therefore 1 kilogram per liter. The density of gold is 19 kilograms per liter. People would say loosely that gold is "nineteen times heavier" than water.

Dipole: A *molecule* whose two ends are slightly charged, one end positively and the other end negatively. The molecule therefore has two electric poles, analogous to the two magnetic poles in a magnet. Water is a common example. The oppositely charged ends of water *molecules* attract each other, making water difficult to boil and vaporize compared with similar liquids.

Electrolyte: A liquid that conducts electricity because it contains electrically charged particles (*ions*). Salt water is the most common electrolyte.

Electromagnetic radiation: Pure energy in wave form, traveling through space at the speed of light. Electromagnetic radiation ranges in energy from low-energy radio waves to microwaves, light rays (both visible and invisible), X rays and high-energy gamma rays.

Electron: A tiny, negatively charged particle. Its native habitat is outside the nucleus of an *atom,* but electrons are easily detached from their atoms and under the influence of a voltage can be made to move through a gas or a metal wire from one place to another.

Equilibrium: A situation in which nothing is changing because all forces are balanced. In some equilibrium situations we may not be able to see any change, but down at the molecular level two opposite processes are taking place at equal rates.

Excitation: An *atom* or *molecule* is said to be excited when it has received some energy in excess of its normal "resting" state. It will usually emit that excess energy within a very short period of time.

Gravitation, or **gravity:** A force of attraction between any two objects that have *mass*. The strength of the force is proportional to the amounts of mass in the objects and becomes weaker the farther apart they are. Earth has a huge mass and is therefore the primary source of gravitational attraction that we normally experience.

Half-life: The amount of time it takes for an amount of a radioactive substance to diminish to one-half of that amount. The amount diminishes because the *atoms* of a radioactive substance are unstable, and are spontaneously converting themselves into different kinds of atoms that are more stable.

Halogen: A family of chemical elements with similar properties. The members of this family are fluorine, chlorine, bromine, iodine and astatine.

Heat: A form of energy that is exhibited by the motion of *atoms* and *molecules.*

Hydrogen bond: A weak attraction between certain *molecules* that contain hydrogen *atoms.* Hydrogen bonds are very important in determining the unique properties of water and many biologically important chemicals, including DNA.

Ion: An *atom* or group of atoms that has acquired an electric charge by gaining or losing one or more *electrons*.

Kinetic energy: The form of energy that a moving object has. Energy of motion.

Mass: The quality of "heaviness" that is possessed by all things, all objects, all matter, all stuff, from subatomic particles to galaxies. All mass exerts a *gravitational* attractive force on all other mass. The effect of Earth's gravitational force on an object is the object's weight.

Molecule: A conglomeration of *atoms,* all bound together. All substances are made of molecules (except a few that are made of unbound atoms). Different substances have different properties because their molecules contain different collections of atoms bound in different arrangements.

Momentum: A measure of how much damage a moving object can do in a collision with another object. Momentum is a combination of the object's *mass* and its speed. The heavier it is and the faster it is moving, the more momentum it has.

Nucleus: The incredibly tiny and incredibly heavy central core of an *atom*. It is thousands of times heavier than all the atom's *electrons* combined.

Photon: A "particle" of light or other *electromagnetic radiation.*

Pressure: The amount of force being applied to an area of surface. All gases exert a pressure on all surfaces that they are in contact with, because their *molecules* are in constant motion and are bombarding the surface.

Quantum: A "piece" of energy. Energy and *momentum* aren't continuous, but exist in tiny, discrete quantities called quanta (plural of quantum).

Refraction: The bending of light or sound waves when they leave one medium (such as glass or air) and enter a different medium in which their speed is different.

Solubility: The quality of dissolving or being dissolved. Chemists use the word to mean the maximum amount of a substance that can be dissolved in a liquid under any particular conditions. The solubility of table salt (sodium chloride) in water at 32 degrees Fahrenheit (0 degrees Celsius) is 357 grams per liter.

Temperature: A number that expresses the average *kinetic energy* of the *molecules* in a substance. The hotter a substance is, the faster its molecules are moving.

Terminal velocity: A fancy expression for final speed. When an object falls through the air from a high place, it will fall faster and faster due to the *acceleration* of *gravity* until the air resistance builds up enough to stop the acceleration, after which the object will fall no faster. It will have reached its terminal velocity.

Vapor pressure: In every solid or liquid substance, but especially in liquids, there is a certain tendency for the *molecules* to become detached from their fellows and go off into the air as a vapor. The strength of that tendency is called the vapor pressure of the substance.

Viscosity: The "thickness" of a liquid; its resistance to flowing freely. As the old saying goes, "Blood is more viscous than water."

Index